Contents

Series editor's preface		7
Preface		9
Acknowledgements		11
SECTION A	**PRINCIPLES**	13
CHAPTER 1	IT into the curriculum	15
CHAPTER 2	Frameworks for IT use in Science and Technology education	23
SECTION B	**PRACTICE**	29
CHAPTER 3	Introduction to the use of IT in school Science and Technology education	31
CHAPTER 4	CAL/tutorial uses: instructional uses of IT in Science and Technology education	37
CHAPTER 5	Simulations	45
CHAPTER 6	Modelling	51
CHAPTER 7	Databases	57
CHAPTER 8	Data-logging	61
CHAPTER 9	Interactive video	69
CHAPTER 10	Word-processing and desk-top publishing	75
CHAPTER 11	Spreadsheets	81
CHAPTER 12	Computers in technology laboratories	83

SECTION C	**ISSUES AND POLICY**	89
CHAPTER 13	Issues to be addressed	91
CHAPTER 14	Issues for the future	99
Bibliography		105
Index		109

Information Technology in Science and Technology Education

DEVELOPING SCIENCE AND TECHNOLOGY EDUCATION

Series Editor: Brian Woolnough
Department of Educational Studies, University of Oxford

Current titles:

John Eggleston: *Teaching Design and Technology*
Jon Scaife and Jerry Wellington: *Information Technology in Science and Technology Education*
Joan Solomon: *Teaching Science, Technology and Society*
Clive Sutton: *Words, Science and Learning*

Titles in preparation include:

David Layton: *Technology's Challenge to Science Education*
Michael Poole: *Beliefs and Values in Science and Technology Education*
Keith Postlethwaite: *Teaching Science to Children with Special Educational Needs*
Michael Reiss: *Science Education for a Pluralist Society*

Information Technology in Science and Technology Education

JON SCAIFE and JERRY WELLINGTON

Open University Press
Buckingham · Philadelphia

Open University Press
Celtic Court
22 Ballmoor
Buckingham
MK18 1XW

and
1900 Frost Road, Suite 101
Bristol, PA 19007, USA

First Published 1993

Copyright © Jon Scaife and Jerry Wellington 1993

All rights reserved. No part of this publication may be
reproduced, stored in a retrieval system or transmitted in
any form or by any means, without written permission from the
publisher.

*A catalogue record of this book is available
from the British Library*

Library of Congress Cataloging-in-Publication Data
Scaife, Jon, 1951–
    Information technology in Science and Technology education/Jon
  Scaife and Jerry Wellington.
    p. cm. – (Developing Science and Technology Education)
    Includes bibliographical references and index.
    ISBN 0-335-09924-6   ISBN 0-335-09919-X (pbk.)
    1. Science–Study and teaching. 2. Technology–Study and
teaching. 3. Information technology. I. Wellington, J.J. (Jerry
J.) II. Title. III. Series.
Q181.S33   1992
507′.8–dc20   92-8265
                                                                                                               CIP

Typeset by Graphicraft Typesetters Limited, Hong Kong
Printed in Great Britain by St Edmundsbury Press,
Bury St Edmunds, Suffolk

Series editor's preface

It may seem surprising that after three decades of curriculum innovation, and with the increasing provision of a centralised National Curriculum, it is felt necessary to produce a series of books which encourage teachers and curriculum developers to continue to rethink how Science and Technology should be taught in schools. But teaching can never be merely the 'delivery' of someone else's 'given' curriculum, it is essentially a personal and professional business in which lively, thinking, enthusiastic teachers continue to analyse their own activities and mediate the curriculum framework to their students. If teachers ever cease to be critical of what they are doing then their teaching, and their students' learning, will become sterile.

There are still important questions which need to be addressed, questions which remain fundamental but the answers to which may vary according to the social conditions and educational priorities at a particular time.

What is the justification for teaching Science and Technology in our schools? For educational or vocational reasons? Providing Science and Technology for all, for future educated citizens, or to provide adequately prepared and motivated students to fulfil the industrial needs of the country? Will the same type of curriculum satisfactorily meet both needs or do we need a differentiated curriculum? In the past it has too readily been assumed that one type of science will meet all needs.

What should be the nature of Science and Technology in schools? It will need to develop both the methods and the content of the subject, the way a scientist or engineer works and the appropriate knowledge and understanding, but what is the relationship between the two? How does the student's explicit knowledge relate to investigational skill? How important is the student's tacit knowledge? In the past the holistic nature of scientific activity and the importance of affective factors such as commitment and enjoyment have been seriously undervalued in relation to the student's success.

And, of particular concern to this series, what is the relationship between Science and Technology? In some countries the scientific nature of technology and the technological aspects of Science make the subjects a natural continuum. In others the curriculum structures have separated the two leaving the teachers to develop appropriate links. Underlying this series is the belief that Science and Technology have an important interdependence and thus many of the books will be appropriate to teachers of both Science and Technology.

The use of information technology (IT) in schools has been subject to much personal advocacy and political expedient. This book, by Jon Scaife and Jerry Wellington, provides a timely, comprehensive and perceptive analysis of the various ways that IT can be used in schools, discussing both the strengths and the weaknesses of different practice. It should be required reading for anyone with the responsibility for some, or all, of computer use in schools.

We hope that this book, and the series as a whole, will help many teachers to develop their Science and Technology education in ways that are both satisfying to themselves and stimulating to their students.

Brian E. Woolnough

Preface

The aim of this book is to provide an introduction to the use of information technology (IT) in Science and Technology education. The book is not a manual or a teacher's guide to using IT, nor is it (at the other extreme) a theoretical tome on the place of IT in Science and Technology education. Our aim has been to integrate the more theoretical frameworks for IT with examples of practice and considerations of policy – so the three themes throughout the book are: principles, practice and policy.

The title of the book, *Information technology in Science and Technology education*, sounds initially like a fairly straightforward pointer to its subject. But a brief reflection on the developments which have occurred in the school Science curriculum and in the understanding of Technology education in recent years reveals that seemingly familiar terms have taken on broader meanings. Rather than offering definitions or idealisations of Science and Technology education, we have tried to represent aspects of these disciplines as we, and others, see them in practice. Thus the Science curriculum to which we refer will include skills and processes as well as facts and concepts, and Earth Science, Astronomy and Materials as well as 'traditional' aspects of Biology, Chemistry and Physics. We have interpreted 'Technology education' as the set of processes which constitute the Design and Technology component of the National Technology Curriculum in England and Wales, namely identifying needs and opportunities, generating a design, planning and making, and evaluating. This scheme, which has existed, at least in spirit for over two decades, is sufficiently broad and flexible as to be unconfined by national boundaries. In addition, we have included in our brief some aspects of the cross-curricular IT component of the National Curriculum in England and Wales which connect with our interpretations of Science or Technology education.

In preparing the book, case studies of IT use in schools were carried out to ensure that the book has at least an element of realism. Some of the schools were recommended to us by teachers and advisers as places where 'things were happening', although none of them would be conceited enough to pretend that they are at the forefront of IT. The cases did show what can be achieved within some fairly difficult constraints, and they also begin to highlight what those constraints, and the barriers to the spread of IT use in Science and Technology education are. Material from the case studies has been integrated into the book at appropriate points, particularly in Section C.

We would like to thank all the teachers and pupils at the schools listed below who allowed us to observe and study their use of IT in the Science and Technology curriculum: Carter Lodge; Nook Lane; Concorde Middle; Wyggeston Queen Elizabeth College. We would also like to acknowledge the assistance we received from staff and pupils at Leysland High School and Stocksbridge Junior School.

Acknowledgements

The following organisations are gratefully acknowledged for supplying photographic material:

Acorn Computers; stills from the interactive videodisc 'Motion: a visual database' by courtesy of Division of Mathematics, Statistics and Physics, Faculty of Built Environment, Science and Technology, and Media Production Division, Faculty of Educational Services, Anglia Polytechnic University; AVP; Educational Electronics; Griffin and George; Lawrence Rogers, Leicester University; Nook Lane Junior School, Sheffield; Philip Harris Ltd; Research Machines Ltd; Stocksbridge Junior School; Valiant Ltd; Woodform Ltd.

SECTION A

PRINCIPLES

CHAPTER 1

IT into the curriculum

Innovation or imposition?

It is now widely accepted that information technology (IT) has an important part to play in enriching Science and Technology education. But the entry and assimilation of IT into the curriculum has had an interesting, if somewhat topsyturvy, recent history. This chapter traces the course of IT's introduction into education since 1980 as a prelude to considering its position and potential in Science and Technology curricula. The purpose of this initial (albeit brief) account is to provide some historical context in which to consider in detail the current and future place of IT in Science and Technology education.

The boom decade

The scale of investment in the microcomputer during the 1980s in the UK was on a level which no other item of educational technology has ever equalled or is ever likely to match. Kenneth Baker, the Minister for Information Technology in 1981, launched the 'Micros in Schools' scheme by claiming that the 'kids of today' should be equipped with 'skills' for the information age analogous to the skills that had gained their ancestors employment:

> I want to try and ensure that the kids of today are trained with the skills that gave their fathers and grandfathers jobs... And that is the reason why we've pushed ahead with computers in schools. I want youngsters, boys and girls leaving school at sixteen, to actually be able to operate a computer.

This clarion call, emphasising the 'vocational' aspect of IT education, is still common, as we will see later.

Baker's words were matched with money. No less than £16 million was provided by the Department of Trade and Industry (DTI) to subsidise the purchase of *British* computers in schools. This sum was exceeded by the Department of Education and Science (DES) supported initiative to promote microelectronics teaching and the use of computers in school education: £23 million launched the Microelectronics Education Programme (MEP) which ran until 1986.

Similar initiatives were occurring elsewhere. It is not the aim of this book to provide a comparative study of IT use but it is worth merely noting the French initiatives of the early 1980s. In 1980, the French Ministries of Education and Industry launched the '10 000 Computers' initiative to provide all upper-secondary schools with *French*-made microcomputers. This was followed, in 1983, by the '100 000 Computers' plan to install that number (a tenfold increase) of computers in schools and to make provision for 'all teachers to learn computer programming' (Dieuzeide, 1987).

Parallel moves were occurring in other countries in the same period, including Australia, New Zealand and the USA (a full review is given in 'Computer Education and the Curriculum', in the *Journal of Curriculum Studies*, 1990, vol. 22, no. 1, 57–76).

Meanwhile, in the UK, other sources of support rose like springs to provide tributaries to the main flow of funds into school computing. The DTI belatedly subsidised educational software to the tune of £3.5 million, having been told repeatedly that, crudely speaking, a computer without software is like a car without petrol. The 1986 modem scheme used another £1 million from the DTI to support the communications facet of IT, hitherto neglected, though the issue of who pays the telephone bill was left untouched. In the same year, the Microelectronics Support Unit (MeSU) was set up with £3 million to carry on the good work of the MEP.

Meanwhile, an educational giant bearing gifts had risen in 1983. The Technical and Vocational Educational Initiative (TVEI) provided manna to computer enthusiasts whose aim was to fill the school with microcomputers and associated peripherals. Though TVEI was in no way meant as a source of financial support for microcomputing, in many lucky schools it was seen that way and, by 1986, some TVEI schools had as many as 50 or 60 micros. A survey in that year showed that TVEI schools had, *on average*, almost twice as many computers as non-TVEI schools (Wellington, 1989a). Finally, in 1988 the DES invested £19 million in educational support grants for IT, £8.5 million of this on hardware.

The main initiatives and landmarks in IT education are summarised in Table 1.1. What were the effects of this 'boom decade'? How far had IT penetrated classroom practice and the school curriculum by the beginning of the 1990s? The next two sections review the impact and influence of IT on school curricula. Purely for the sake of simplicity, the primary and secondary sectors are considered separately before a framework is presented for assessing the curricular impact of IT.

IT in the secondary curriculum: one down, two across

In the UK, prior to the DTI 'Micros in Schools' scheme, Computer Studies as a secondary school subject was on the fringe of the curriculum, and as an examination subject its entry figures were at a similar level to Spanish, Geology and Music. By 1984, examination entries had virtually tripled to place the subject firmly in the mainstream of the secondary curriculum.

This strategy for introducing computer education into the secondary school curriculum was, however, already beginning to be questioned. Firstly, the subject Computer Studies was rapidly becoming the 'domain of the boys' with a boy:girl ratio of 2.4:1 at the 1984 O level entry, a male bias exceeded only by Physics with a male dominance of 2.7:1. Secondly, questions were raised about the content of Computer Studies courses. The course content often involved topics such as the history of computing, the representation of numbers and characters in binary notation, programming in BASIC, and the study of Logic, which the subject's critics considered neither educationally worthwhile nor vocationally relevant. Finally, the subject at school level received criticism from the influential Alvey Report and from Universities and Polytechnics in the UK. The latter group failed to favour that area at school level, preferring traditional subjects at O and A level. The Alvey Report (1982: 62) went further, suggesting that school computer education of the wrong kind (and the use of home micros) might actually do harm, and by implication, prejudice a student's chances of entering higher education:

> it is no good just providing schools with microcomputers. This will merely produce a generation of poor BASIC programmers. Universities in fact are having to give remedial education to entrants with A Level Computer Science. Uncorrected, the explosion in home computing with its 1950s and '60s programming style will make this problem even worse.

Alongside the almost exponential growth of Computer Studies there lay a second initiative which fought against that growth: the MEP which ran from 1980 to 1986. One of its major aims was to encourage the use of computers as aids to teaching and learning *across* the school curriculum.

At secondary level, the first main deviation from Computer Studies as an examination subject was

IT INTO THE CURRICULUM

Table 1.1 Some landmarks in IT education (1980–8)

Date	Title	Description
1981–4	Micros in School Programme	£16 m. subsidy from DTI for purchase of microcomputers in schools
1980–6	Microelectronics Education Programme (MEP)	DES supported initiative (£23 m.) to promote microelectronics teaching and use of computers in school education
1981	ITeCs (Information Technology Centres)	Government initiative to train young people in IT skills; 175 ITeCs now operating, ca. 6 000 trainees
1982	Information Technology Year (IT) 82	Government programme to increase national awareness of IT
1982	YTS	Youth Training Scheme
1982	IT initiative launched for higher education	Designed to increase number of places in higher education and postgraduate courses related to IT, 1983–6
1982	'New Blood' Initiative	Money provided for higher education to recruit lecturers/researchers in IT (70 posts)
1983	TVEI	Technical and Vocational Educational Initiative
1984	Information Technology in Further Education	Government grant designed to provide vocational students with an education taking account of the industrial and commercial applications of IT
1985	Engineering and Technology Initiative	£43 m. funding to increase the number of places in engineering and technology in higher education by around 5 000 by 1990
1985	IT SA (IT skills agency)	Agency set up by industry to monitor skill shortages in IT and encourage collaboration of industry, government and education
1985–6	DTI software subsidy	£3.5 m. provided to DTI to subsidise purchase of educational software in schools
1986	Modem Scheme	DTI subsidy of £1 m. to enable schools to purchase a modem for their micro, allowing links between computer systems
1986–7	Microelectronics Support Unit (MESU)	Set up with £3 m. funding for 1986–7 to carry on the work of the MEP
1986	TVEI extension announced	White Paper, *Working Together – Education and Training*, announced national extension of TVEI programme, with 'average annual expenditure of £90 m.' over next 10 years
1987	DES support for IT in schools and further education	Kenneth Baker announced Educational Support Grants of £19 m. for the expansion of IT in the schools and £4.8 m. for IT in non-advanced FTE.

to provide 'Computer Appreciation' courses for all pupils, often at second- or third-year level. However, a further stage (valued so highly by MEP) was confronted by obstacles both physical and mental. This was the stage of introducing computers across the curriculum into separate subjects, e.g. computer assisted learning (CAL) across the secondary curriculum. The physical obstacle, of course, had been the creation of 'computer rooms' having anything between 10 and 20 computers and often part of the domain of the Computer Studies and Maths teachers.

Therein also lay the mental obstacle. Computers became widely seen, with some notable exceptions, as the province of the Maths/Computer Studies departments, kept under lock and key in a computer room which often had to be booked well in advance, and in many cases contained micros linked or networked together. This prevented their 'physical diffusion' into the fabric of the school, and their 'mental diffusion' into the curriculum planning and classroom practice of other teachers.

An additional problem in the UK had been the lack of *technical* support. If computers are to diffuse through a school into the rooms and practices of a wide range of teachers, technical assistance is needed. This may be provided, in exceptional schools, by the dedicated and overworked 'computer teacher' – but in the majority of cases the safe option was taken. Micros remained in the computer room under a watchful eye away from the 'incompetence at large'.

The alternative is for the school 'computer teacher' to act as a support technician to teachers in other subjects. This is yet another role for the teacher in charge of IT. That person's role has evolved from a subject pioneer in developing, learning and teaching a totally new subject to a provider of in-service training. For IT education to diffuse, that teacher is also forced to provide technical support for other staff in a range of subjects, in addition to providing further in-service training, and in some cases suggesting and evaluating suitable software across the curriculum. This triple role of technician, in-service training and software provider cannot be sustained by one individual unless he or she is given the time and freedom to do it. Without the provision of co-ordination and support for IT education, the goal of 'IT across the curriculum' is unlikely to be attained.

In summary, then, it became clear during the 1980s that IT was making an impact on the secondary curriculum in three very different but often competing ways:

1 by offering 'Computer Studies' as a subject in itself which could be chosen as an option, leading to an examination;
2 by creating computer awareness or appreciation courses for all pupils either as a topic or module in itself, or as an ingredient of existing subjects across the curriculum, e.g. History, Mathematics, English, etc.; and
3 by introducing computer assisted learning as an element in traditional school subjects.

At first sight these perpendicular approaches might seem to be perfectly capable of co-existing. However, it became clear that the initial concentration on the vertical approach, i.e. the addition of Computer Studies and subsequently IT as examinable subjects, did have certain unwanted side-effects. It emerged that an excessive concentration on computing courses for a minority severely restricted the aim of IT awareness and familiarity for all, and the use of CAL in other subjects. This would not occur in an ideal world of course, but resource and in-service training provision were never large enough to achieve every aim. In short, where examination courses dominated, IT across the curriculum suffered.

These side-effects have lingered on into the 1990s which is why an historical sketch of this kind is so important for understanding the position of IT in Science and Technology education at present. Many of the constraints identified by teachers, which surface in the case-study evidence presented later, are a direct result of the policies and trends of the 1980s. These perceived constraints include the lack of easy access to IT facilities, unsuitable in-service training, the restrictions imposed by the 'computer room', inappropriate hardware and software, and so on. These issues are taken up in later chapters.

IT in the primary curriculum: forwards or backwards?

As a curriculum innovation, the introduction of microcomputing into the primary curriculum has been unique in its fivefold combination of central cash support, parental backing, motivation of children, commercial hype, and potential for home/school liaison. These five factors alone would seem to make its impact upon, and success within, school life a virtual certainty. Yet doubts were increasingly being cast in the mid-1980s by participants at all levels over the effectiveness of computer education and its ability to enhance 'good practice'. Chandler (1984), for example, in discussing the 'young learner', prefaces his book by asserting: 'The microcomputer is a tool of awesome potency which is making it possible for educational practice to take a giant step backwards into the nineteenth century.' Similar cynicism is expressed by Weizenbaum in a 'telephone interview' published in full by Wellington (1985):

> the introduction of microcomputers into primary and secondary schools is basically a mistake based on very false assumptions. Too often, the computer is used in schools, as it is used in other social establishments, as a quick technological fix.

By 1985 it had almost become fashionable to pour scorn on the 'Micros in Schools' scheme, the software produced for computer education, and even the hardware employed. Such scepticism, for example, formed the underlying theme of the BBC series *The Learning Machine* (O'Shea, 1985) on educational computing and an important book by Self (1984) on educational software.

In the space of five years, then, the optimism and missionary zeal of 1981 was being replaced by growing criticism of the hardware and software of computer education. But in the meantime, little study had been made of the impact of educational computing at grassroots' level. How far had this highly-funded innovation penetrated the practices, or even the philosophy of classroom teachers? How were teachers using the computer in their classrooms? What barriers or obstacles did they perceive to the permeation of computer use into their own practices and curricula? How far had the multi-million pound initiatives of central government cascaded, or trickled, down into classroom work?

In primary education two studies in 1986 indicated that computer use in primary schools was not widespread and was rarely integrated into 'good classroom practice' (Bleach, 1986; Ellam and Wellington, 1986). Ellam and Wellington suggested that the 'human factor' in introducing educational computing was often neglected, with an over-concentration on providing hardware and software. The DTI initiative to provide a £3.5 million subsidy for schools to purchase software did little to overcome human barriers to innovation.

What were these barriers to innovation? The fear of technology, innovation and change is a key factor. Even people who work regularly with computers have a love–hate relationship with them. Those who have never worked with new technology are likely to have a similarly ambivalent attitude – a mixture of anticipation and yet anxiety. This will be particularly true of the classroom teacher who views the introduction of new technology as perhaps a threat, a challenge, a Trojan Horse or yet another innovation with which to get to grips. This complex chemistry of teacher attitudes has probably been the major barrier to success. Attempts to 'change' these attitudes by intensive in-service training programmes have rarely succeeded – simply sending teachers out on to a short sharp course and expecting them to come back and influence their colleagues is far too simplistic.

Fitting new technology on to daily working practices in the classroom is a problem – fitting the innovation into the primary curriculum is often equally difficult. How many innovations in IT were designed to solve a particular curriculum problem or to meet a particular teacher-need? So many innovations have been rightly labelled as solutions in search of a problem.

Thus in many ways the important route to success in developing IT in the primary curriculum was not initially followed – the education must come first, the technology second.

Again, these issues emerged in the case-study evidence from the primary sector presented later. Teachers' attitudes to IT, scepticism about its relationship to 'good primary practice' and discontent with in-service provision are still widespread in the 1990s. These perceptions linger in the minds of teachers, many of whom taught through the IT innovations of the 1980s, and have a deep effect on current practice.

The curricular impact of IT: three pressures and three waves

So far the discussion of IT in the curriculum has been a mixture of historical account and brief discussion. This section offers two frameworks for considering the impact of IT on the curriculum. Both are useful in considering (both here and in later chapters) the role of IT in Science and Technology education.

Three pressures

Cerych (1985) points out that the introduction of IT into the curriculum has been unique in involving a variety of pressures or influences exerted upon it. He distinguishes three factors as key agencies in the education–IT 'interface' (1985: 225). He refers to these as 'pedagogical', 'sociological' and 'economic':

- Firstly, IT has entered education as a new pedagogic tool, fundamentally different from tools of the past because of its capability to be *interactive*. As a result, IT has been widely pushed as a learning tool, particularly in the primary curriculum, since it can involve active and enjoyable participation.
- Secondly, the introduction of IT into education has been accompanied by *sociological pressure* – from parents, from local authorities, from successive governments, from European and international organisations.
- Finally, there has been huge *economic pressure* behind the introduction of new information technology into education because IT is not just an education tool (as were, for example, the programmed learning machines of an earlier decade). IT is now pervasive in all economic sectors. The pressure has come from statements on the 'needs of industry', skill shortage and on the 'growing demand for IT skills'. This latter pressure ('economic') has been particularly strong in the secondary sector where the *vocational significance* of IT has always been stressed strongly. Whether or not this emphasis is justified has recently been questioned (Wellington, 1989a).

Three waves

In addition to these unique pressures, three 'waves' (Sendov, 1986) can be seen in the development of IT education in the curriculum. In the first wave, computers were introduced into schools virtually as a new educational facility, almost in the manner of the overhead projector, the tape recorder or the film projector (Sendov, 1986). In many cases they were literally 'dropped on school doorsteps' (Steele and Wellington, 1985a) courtesy of the DTI, or the equivalent body in other countries.

In this first wave, the computer remained very much an appendage to education. As a result, it became an object of study in its own right. In the second wave, the value of the computer and more generally IT as an educational resource, begins to be appreciated and developed. In secondary schools the use of the computer spreads into existing disciplines – more and more teachers begin to see it as a valuable educational resource with great potential for their own subject. IT is seen as cross-curricular rather than the province of the expert. Evidence suggests (Wellington, 1989a), that some secondary schools are firmly into the second wave though diffusion has yet to be completed.

The third wave, which is as yet largely hypothetical, occurs when IT influences the content and the aims of education itself, as well as the method and the system of teaching. There are signs of the beginning of this wave in current developments in data logging which are described later. Sendov (1986) argued that this third wave may occur with the 'mass presence of the computer in the social environment'. Others argue that it may happen with the advent of powerful computers and information systems (Mackintosh, 1986) and with them, powerful ideas (Papert, 1980). What would occur in a third wave is a re-appraisal of the nature and aims of separate school disciplines in the context of powerful information technology systems and new infrastructures.

IT in the school curriculum: where are we now?

What effect have these pressures had on IT in education and where do we stand in relation to Sendov's three waves? In short, what is the current position of IT in the curriculum?

Despite the huge level of funding for IT in the curriculum, teachers still see *shortage of funds* as a major obstacle to its development. This is particularly true in the primary sector, very much the poor relation to secondary in IT terms. We will use the term 'computer access factor' (CAF) to denote the ratio in a school of the number of pupils to the number of microcomputers. The figure improved dramatically from 107:1 in 1985 (DES, 1986) to 69:1 in 1989 in primary schools. An impressive step forward perhaps, but poor in comparison with the CAF in the secondary sector – from 60:1 in 1985 to 28:1 in 1989. Thus the CAF is more than twice as healthy in UK secondary schools than in primary.

Why should this situation have occurred? We would speculate that the *economic* curriculum pressure on IT – in which the perceived vocational significance of computers plays a major role – has been the greatest of the three pressures defined earlier. The *pedagogic* pressure of IT in the curriculum has grown – far more teachers now perceive the computer as a valuable learning resource or vehicle than previously (Macdonald and Wellington, 1989). But this pressure has been largely confined to the primary *teachers* working with IT in the classroom – the policy-makers who make the decisions seem to dance to a different tune. Thus the perceived goal of many primary teachers is still to achieve the target of having *one microcomputer system in every classroom*. This target is still a long way off in the UK and in other countries. The more ambitious target of a 1:6 ratio in primary schools suggested by the *POST Report* of 1991 will be seen by many teachers as simply pie-in-the-sky (*POST Report*, 1991).

The three curriculum pressures outlined earlier thus exist within a complex system of decision-making. A naive view that all pressures act equally cannot be taken. Pedagogic pressure appears to be in a different league from the others.

An interesting feature in the primary sector has been the potent but largely hidden *sociological pressure* on IT in the curriculum from parents. The most important source of money for computing in the primary sector comes from school-raised funds, and within that category almost three-quarters of the money comes from parent–teacher associations of some kind. This is in sharp contrast to industry support for microcomputing in primary schools, which is so small as to be negligible (Macdonald and Wellington, 1989, B28). Our general point is that the three pressures on IT in the curriculum – pedagogic, sociological and economic – operate in different ways and with different strengths, in different sectors of education. Quite simply, the influence of pedagogic pressure in the primary sector is small in comparison to the economic and vocational pressures in the secondary curriculum. This accounts for the major differences in funding and resources between the two areas. Without the third pressure – sociological, principally from parents – that difference would be even greater.

The notion of three *waves* of IT in education provides a second perspective. Many schools are now firmly into the second wave in which the computer is viewed as a learning tool, its use is integrated into existing subjects and a more critical view is taken of the vocational value of school computer education.

What of the third wave? The influence of a vertical, subject-based National Curriculum is a powerful force preventing the emergence of this third stage. Information technology has not yet had any significant effect in changing the dominant view of the *secondary* curriculum based on forms of knowledge, separate subjects and compartmentalised disciplines. In short, the effect of IT upon the content and structure of the curriculum has been negligible.

Optimistic opponents of the subject-based curriculum will therefore hope that the future progress of IT in both society and education may lead us to examine not only *how* we teach but *what* we teach. At first IT may bring into question the nature, aims and content within separate subject areas such as History, Geography, Science and Mathematics. Then, ironically, the diffusion of IT across the existing curriculum may actually undermine the structure of that curriculum and bring into question the wisdom of teaching separate subjects or individual disciplines. This may entail a more 'horizontal' view of the whole secondary curriculum rather than the existing 'vertical' view of discrete subject domains (Wellington, 1985). Ironically, the greatest influence of IT on education may be to make the curriculum in the secondary sector more closely resemble that of the primary school.

CHAPTER 2

Frameworks for IT use in Science and Technology education

Chapter 1 examined the introduction of IT into the curriculum as a prelude to considering its use in Science and Technology education. This chapter discusses the potential of IT in those areas by considering the technology itself, classifications of learning which can result from IT use, and finally (very briefly) possible justifications for the use of computers in Science and Technology education.

What are computers good at?

One of the starting points for considering the applications of IT in education is to consider carefully just what computer systems can and cannot do. This may seem a very *technology-centred* starting point but a more *learner-centred* approach will be introduced in the next section.

Generally speaking, modern computer systems are good at:

- collecting and storing large amounts of data
- performing complex calculations on stored data rapidly
- processing large amounts of data and displaying it in a variety of formats

These capabilities all have direct relevance to the process of education, and at the same time raise important issues for education. One issue, which is introduced here simply by way of example and will re-appear later in other chapters, concerns the use of computers as labour-saving devices. As listed above, computers can collect data at a rapid rate and perform calculations extremely quickly. But the question arises: should the computer (in an *educational* context), be used to collect, process and display rather than the learner? In other words does the use of a *computer* in saving labour take away an important educational experience for the *learner*? A similar issue appears in the use of computers and electronic calculators to perform complex calculations rapidly. This may be desirable in some learning situations, e.g. if the performance of a tedious calculation by human means actually impedes or clutters up a learning process. But it can also be argued (and often is by people perhaps wrongly labelled 'traditionalists') that the ability to perform complex calculations rapidly should be one of the *aims* of education, not something to be replaced by it.

In our view, the distinction between what counts as *authentic* (i.e. desirable and purposeful) and *inauthentic* (i.e. unnecessary and irrelevant) labour in the learning process is a central one in considering the use of IT in education. The notions of 'inauthentic' and 'authentic' labour will be revisited in later chapters.

Examine the above suggestions of 'what computers are good at'. Could there be any missing from that list? Is the distinction between 'authentic' and 'inauthentic' labour in the learning process an important one?

It is also worth noting that computers do exactly what they are instructed to do, very quickly and consistently. On the one hand, this means that they are not (or at least not yet*) capable of making autonomous or independent judgments, or personal interpretations. However it is also the case that they do not become tired, bored, hungry, irritable, angry or impatient, or liable to error. This may place them at an advantage in some situations as compared to teachers! It has been said that *one* of the reasons why children appear to enjoy learning with computers is precisely because of these impersonal, inhuman qualities.

One final point on the abilities of computers is worth stressing. Computers can, in a sense, speed up, or slow down, reality. As Kahn (1985) puts it:

> they operate outside the viscous flow of time in which humans perform tasks.

This is an important point which will be elaborated upon and discussed when the use of computer *simulations* in education is considered in Chapter 5. Computer simulations do, in some way, distort time ... and perhaps reality.

Types of IT use in education

We should now look at IT from the learner's perspective. There are a number of ways of classifying IT use in education. In our view the most useful classification dates back to 1977 and was produced by Kemmis, Atkin and Wright (1977). This seminal paper identified four paradigms by which students learn through the use of IT (a 'paradigm' can be defined as a 'pattern, example or model'). They are:

- the instructional paradigm
- the revelatory paradigm
- the conjectural paradigm
- the emancipatory paradigm

We will consider each one briefly in turn but

---

* The advent of artificial intelligence and so-called expert systems does give rise to some debate on this point. For example, it might be argued that some intelligent, knowledge-based systems (IKBS) are capable of making judgments. But are they 'autonomous'?

further reading is necessary to consider them fully and reflectively. (See, for example, Rushby, 1979; Wellington, 1985; Blease, 1986; Sewell, 1990.)

The instructional paradigm

The overall aim in this paradigm is to teach a learner a given piece of subject matter, or to impart a specific skill. It involves breaking a learning task into a series of sub-tasks each with its own stated prerequisites and objectives. These separate tasks are then structured and sequenced to form a coherent whole.

Computer-assisted learning (CAL) of this type is now given names like 'skill and drill', 'drill and practice' and 'instructional dialogue'. Perhaps its main problem is that in the early 1980s some teachers and others involved in education saw it as the dominant paradigm in CAL. This probably resulted in their poor perception of educational programs and the belief that microcomputers were a 'passing fad in education' like the programmed learning machines of the 1960s.

Fortunately, the increasing prevalence of the other paradigms has made this view untenable.

The revelatory paradigm

The second type of IT use involves guiding a student through a process of learning by discovery. The subject matter and its underlying model or theory are gradually 'revealed' to the student as he or she uses the program.

In contrast to the instructional form, where the computer presents the subject matter and controls the student's progress 'through' it, in revelatory CAL, 'the computer acts as a mediator between the student and a hidden model' of some situation (Rushby, 1979:28). The revelatory paradigm is now exemplified in educational programs by numerous simulations, of various types: *real* (e.g. an industrial process), *historical* (e.g. empathising with a historic event), *theoretical* (e.g. the particle theory of matter), or even *imaginary* (e.g. a city of the future).

FRAMEWORKS FOR IT USE IN SCIENCE AND TECHNOLOGY EDUCATION

The conjectural paradigm

This third category involves increasing control by the student over the computer by allowing students to manipulate and test their own ideas and hypotheses, e.g. by allowing modelling.

Modelling must be distinguished from *simulation*. Every simulation involves using a simplified representation, i.e. a model, of some situation, but in a simulation the model is ready-created by the programmer. The user can then alter and experiment with the external conditions and variables affecting the model, but cannot tamper with the model itself, i.e. internal conditions. In modelling, however, the user creates a model of the situation, and may then go on to test it, for example by seeing how well it represents and predicts reality.

The potential of model building and model testing has hardly been tapped in educational computing, except perhaps in Science and Technology education (see Chapter 6). Creating and testing models may slot most easily into Science and Technology courses. A model can be formed of some physical phenomenon, e.g. the expansion of a liquid or the motion of a projectile. The patterns predicted by the model could then be compared, say, with the results of an experiment. Clearly, this involves far more control by the learner over the computer. A similar modelling exercise could be used in history, e.g. by studying data in a local census, searching for patterns and forming hypotheses. These hypotheses could then be tested by studying further data, and searching for new evidence in their support.

Encouraging pupils to create, use and test their own models will have great educational value – unfortunately the educational software to enable this is likely to be time-consuming and expensive to produce.

The emancipatory paradigm

The fourth and final paradigm involves using a computer as a labour-saving device – a tool which relieves mental drudgery. As such, it can be used for calculating, for tabulating data, for statistical analysis, or even for drawing graphs. In this type of CAL, the learner uses the computer as and when he or she wants to as an unintelligent, tedium-relieving slave in aiding his or her learning task. This relies on the distinction made earlier between *authentic* labour and *inauthentic* labour. The authentic labour is the central, indispensable part of the learning task. The inauthentic labour is not an integral part, nor is it valued for its own sake, but is still necessary, e.g. doing endless calculations, searching though a filing cabinet, sorting information into alphabetical order, making a bibliographic search, etc. The distinction is not always an easy one to make. Doing calculations, as mentioned earlier, may be seen as a worthwhile exercise in itself. But where the distinction can be made the computer *can* be seen as a useful tool, e.g. in handling information in a history lesson.

This fourth type of CAL is perhaps unique in two ways: firstly, it uses the computer purely as a tool for the learner's convenience, to be used when and where it is needed; secondly, the computer is only partly involved in the learning process, i.e. to take over the 'inauthentic' part of the learning task.

These four proposed paradigms are summarised in Table 2.1. The focus (or locus) of control is shown changing from the computer to the learner, as the type of IT use shifts along the paradigms. Look closely at the table. How useful are these paradigms? Are there any educational programs, or types of IT use which do not fit into any of these paradigms? Have they stood the 'test of time' since 1977?

Other classifications of IT in education

There are other ways of classifying educational software, and types of IT use in education, which should be considered. A summary is given in Table 2.2. Chandler (1984), for example, suggested six models of CAL which depend on the locus of control. In some ways they are similar to the four paradigms but perhaps reflect more accurately the

Table 2.1 Categories of learning with IT

1 Instructional	2 Revelatory (Simulation)	3 Conjectural (Modelling)	4 Emancipatory (Labour-saving)
Drill and practice Programmed learning, e.g. structured Q & A dialogue in a definite sequence Learner is led by computer	*Of real situations:* Trying out an existing model Varying external conditions Discovering the nature of a model, i.e. guided discovery learning *Of imaginary situations:* Games of adventure, logic and skill Educational games	Making and testing a model of reality Testing ideas and hypotheses Drawing conclusions and discovering patterns from a set of data, e.g. historical models	Computer as a labour – saving device, e.g. calculating, drawing graphs, capturing data, statistical analysis, filing data, retrieving information, text processing
COMPUTER IN CONTROL Subject-centred Content-laden Computer programming children	←――――――――――→		STUDENT IN CONTROL Learner-centred Content-free Children programming computer

wide range of educational software now available. Maddison (1982) provided an interesting distinction between programs whose 'internal workings' are clear or transparent ('glass boxes'), and those which give no indication at all of their internal structure ('black boxes'). He suggests that the difference between programs which are 'glass boxes' and those which are 'black boxes' will affect the way they are used by teachers in the classroom.

Finally, Gray (1984) offered a succinct but quite interesting classification of software apparently based on his own interest in ornithology. Imagine a bird-watcher in a field faced with a choice of telescopes for observing. One type of telescope has a 'fixed focus', used for looking at one bird only in one place (cf. tutorial/drill-and-practice programs). A second type is the 'pre-focused' telescope which can swing round to look at other birds and can pan across the scene (cf. the revelatory paradigm). The third telescope is the 'variable focus', and can focus on anything at all that the user chooses (cf. content-free, tool software).

Gray's ornithological analogy is a useful tool for thought but does not allow the possibility that the area being observed (the terrain) could be chosen and determined by the user. With the locus of control firmly in the user's hands the choice of both terrain and telescope is open. This is surely the kind of IT which Science and Technology education should be working towards?

Justifying the use of IT in Science and Technology education

One way of examining or justifying the use of IT in education is to divide possible justifications of IT use into two categories – educational justifications and vocational justifications.

FRAMEWORKS FOR IT USE IN SCIENCE AND TECHNOLOGY EDUCATION

Table 2.2 Classifying educational software: a summary

1 Kemmis et al. (1977): Four educational paradigms

Paradigm	Features/examples
instructional	Skinnerian; tutorial; drill and practice
revelatory	simulations
conjectural	databases, modelling, LOGO, spread sheets
emancipatory	databases, data-handlers, data-loggers (removing inauthentic labour)

2 Chandler (1984): Six models, depending on the locus of control

Model	User as ...	Examples
Hospital (program in control)	patient	tutorial, drill-and-practice
Funfair	emulator	games
Drama	role-player	adventure games, imaginary world programs
Laboratory	tester	simulations
Resource-centre	artist/researcher	word-processing databases, DTP
Workshop (user in control)	inventor	programming, e.g. with LOGO

3 Maddison (1982): Clarity of structure

Black boxes	Glass boxes
... give no indication of their internal structure	... relatively transparent

4 Gray (1984): Change of focus

Fixed focus
Pre-focused
Variable focus

Educational justifications

Can the use of IT be justified on *educational* grounds? For example:

- does it enhance the learning of certain skills?
- does it promote higher-level thinking by freeing the learner from low-level tasks?
- does it help to develop problem-solving abilities?
- generally, does IT enhance and enrich educational experiences for children?

These are all questions to consider in evaluating IT on educational grounds and should be applied during the chapters which follow.

Vocational justifications

The use of IT in education has often been justified on *vocational* grounds, i.e. in preparing pupils for the 'computer-oriented' society which awaits them. This justification is often given by those who argue that one of the functions of schooling is to provide for, and meet, the needs of industry. Can the use of IT in education be justified for its vocational impact, e.g. in preparing the 'workforce for a technological future'? This question needs critical examination. It cannot be fully discussed here but has been discussed with healthy scepticism, by (for example) Robins and Webster (1989) who talk of 'the technical fix', Roszak (1986) who attacks the 'cult of information' and by Wellington (1989a) who analysed the links between education and employment in terms of IT capabilities. The vocational aspect of computer use should be borne in mind when reading subsequent chapters.

Finally, it should be noted that there may be other justifications for the use of IT in education, and these alternative justifications are discussed in Section B.

SECTION B
PRACTICE

CHAPTER 3

Introduction to the use of IT in school Science and Technology education

Introduction

This section of the book will look at ways in which IT is currently being used in Science and Technology education in schools. We will consider not only the uses which are actually occurring, but also the potential uses of IT which are in many cases already technically possible and available, but for many reasons are not in fact taken up. This will include, for example, interactive video.

Section A provided frameworks for considering IT uses in Science and Technology which we feel to be useful in looking at specific uses of IT in schools. These frameworks will be used throughout this section in discussing classroom activities involving IT. Different categories of use will be presented in different chapters. But the other framework which needs to be borne in mind in school use is the *curriculum* framework, i.e. the guidelines laid down to schools which in practice determine what is taught, and to a large extent *when* and *how*. The discussion in Section A on the introduction of IT into the curriculum gave the general context – here we will present examples of more specific curriculum guidance given to Science and Technology teachers.

IT activities in schools: curriculum frameworks

In the UK, curriculum is at least partly determined by the National Curriculum Council, although the extent to which the actual process of teaching is constrained by it is the subject of debate. In Science, for example, one of the 'profile components' contained the statement that pupils (DES, 1988)

> should develop the intellectual and practical skills that will allow them to develop a fuller understanding of scientific phenomena and the procedures of scientific exploration and investigation.

This had direct implications for the use of IT – how can pupils develop these practical skills and understanding of scientific procedures if they do not use, for example, data-logging equipment? The same document stated that

> pupils should develop their knowledge and understanding of information transfer and microelectronics.

This was to become, at a later stage, the basis for the notorious AT12 which formed one of the statutory attainment targets in the first version of the National Science Curriculum. Very soon afterwards the NCC consultation report on Technology (NCC, 1989a) stated that (NCC, 1989b)

> pupils should be able to use IT to communicate and handle real information; design, develop, explore and evaluate models of real or imaginary situations; and measure physical quantities and control movement.

Again this has implications for classroom uses of IT such as modelling, the use of simulations and of course uses of IT which overlap with science education such as data-logging.

It is not our intention here to examine the National Curriculum for England and Wales in great detail, but it is worth making explicit from the relevant documents the implied uses of IT and also the overlaps between Science and Technology, especially as these uses and overlaps are likely to be similar to those in other education systems.

Science/Technology overlaps

The statutory guidance for the Technology National Curriculum for England and Wales (i.e. the ATs and the programmes of study) contain many statements which form an important part of Technology education but which would fit equally well into Science education. It is not our intention to consider every instance but to select a few examples which illustrate the importance of liaison and the necessity for cross-curricular approaches to IT.

1 In level 3 of the statements of attainment it is said that pupils should be able to 'collect information and enter it into a database'. The suggested example is to 'enter data recording the birds using the school bird table, check the data and retrieve it to compare the numbers and types of birds on different days'. This activity, which might be carried out by 7-year-olds, would link closely with many of the activities forming the Exploration of Science component in the Science curriculum, e.g. looking for patterns and forming hypotheses (it is worthy of note, however, that the levels in the Science and Technology guidance do not always coincide).

A similar example from the Design and Technology component (AT1, level 5) states that pupils should 'use information to compile a database on adults' eating habits'. Another database example appears in level 7. Clearly, the use of databases is an activity which will occur across a wide range of age-groups. This, incidentally, raises the issue of progression in the use of IT which is discussed in Chapter 13. Chapter 7 in this section is devoted to the use of databases.

2 At level 5, pupils are expected to use IT to 'explore patterns and relationships, and to form and test simple hypotheses'. For example, this can be done by 'using a simulation to explore how the populations of predator and prey species fluctuate, and suggest when a predator is most active'. Once again this overlaps with both the Exploration of Science component and the *knowledge and understanding* area of Science. Chapter 5 in this section is devoted to simulations and their use.

3 At a later age, level 7, pupils of perhaps 14 to 16 years are expected to 'design, use and construct a computer model of a situation or process'. The example given is a model of a queue in a supermarket in which pupils should be able to vary the service time, the number of customers and the number of check-outs – no small task! This links with the use of modelling in Science, both as a scientific procedure and as an element of the study of the nature of Science. Chapter 6 in this section considers modelling in some detail, as an aspect of both Technology and Science education.

4 At the same level of the Technology guidance, pupils are expected to understand that 'the results of experiments can be obtained over long or short periods, or at a distance, using data-logging'. One example given is to use IT to measure the acceleration of a model car as it travels down a runway; a later example (level 8) is to use monitoring equipment to record environmental change. The connection with Science education is clear. The use of data-logging in both Science and Technology education is examined later.

5 In another technology attainment target (AT1, level 6) pupils are expected to 'consider the opportunity for newsletter design arising from desk-top publishing'. In AT2, level 8, pupils are expected to 'use computer aided design, image generation and DTP techniques'. The use of word processors permeates both the Science and Technology curricula, and Chapter 10 in

INTRODUCTION TO THE USE OF IT

Fig. 3.1 a, b, c IT in use in Science and Technology education

this section considers both DTP and word-processing.

6 Examples similar to those given above abound in the programmes of study which involve investigations using databases, monitoring the dampness of soil, and the use of a model to study the development of pond algae in different conditions. At the penultimate level (level 9), pupils should be taught to assess how accurately a model reflects reality – examples given being 'a program to trace the trajectory of a tennis ball', and a spreadsheet 'to anticipate trends in predator/prey population'. Again, spreadsheets will be considered later in this section.

The above examples are given simply to show the many areas of overlap between the IT capability section of the National Technology Curriculum and Science education (whatever the curriculum). The examples are not intended as an illustration of any special or unique relationship between Science and Technology – examples could equally well be used to show the vital links between Technology and other areas. They are used in this context to show the common elements of IT use between the two areas which will be examined in detail in subsequent chapters.

It seems then that the key areas of match or overlap between Science education and the IT capability element of Technology are:

- the use of databases
- the use of simulations and (at a higher level) the use of computer modelling
- the use of data-logging or monitoring
- the use of spreadsheets in collecting, comparing and studying results
- the use of general (or 'generic') packages for, say, word-processing or desk-top publishing

These uses of IT will all be discussed using specific examples. In addition, the use of interactive video, although admittedly uncommon in schools, will be reviewed, and also the use of straightforward tutorial style CAL which is certainly more common.

It is interesting to note that all the above uses fit into the frameworks put forward in section A, especially the classification of *instructional*, *revelatory*, *conjectural* and *emancipatory* uses of IT (see p. 26). Tutorial CAL is instructional. Simulations clearly involve revelatory use of software. Databases and data-logging for example can be classed as emancipatory uses of IT, though in some ways also as revelatory and conjectural uses. Spreadsheets are emancipatory but in some contexts involve conjectural use in the same way as modelling programs. Word-processing and desktop publishing programs are obviously emancipatory, but there is far more to their use than merely 'labour saving' – in some ways they fit least easily into the four paradigms. These issues will be followed up in later chapters.

Hardware used in Science and Technology education

We will now give an overview of the hardware which has been, and is being, used in Science and Technology education. This overview is necessary simply because the hardware itself is, and has been, an important determinant of IT use in education.

Microcomputer systems used in schools in the early 1980s generally had tiny memories, e.g. 4K or 8K of RAM, limited storage (cassette tape), poor screen resolution, monochrome display, non-standard keyboards or keypads, and severely limited facilities for connecting to the outside world – typically the only peripheral device which could be connected was a printer. None the less, this still allowed various IT activities to take place, in Science perhaps more readily than in Technology. In particular, tutorial programs, basic simulations and some data-logging (using the printer port) were possible. Other activities were ruled out, principally because of limited memory and storage, or lack of applications software such as word-processing programs.

Looking back, this technology appears so primitive that it is no wonder that the teachers who persevered with it were viewed as dedicated hobbyists. Interestingly, shades of this amateurish view of IT users still linger on. While this dedication was admirable, it may have helped to alienate

INTRODUCTION TO THE USE OF IT

Fig. 3.2 A purpose built trolley for a mobile work-station

Fig. 3.3 The WIMP environment

the majority of Science and Technology teachers from becoming involved in IT. In our opinion it is only very recently (certainly in the UK) that hardware has started to appear in schools which has been designed primarily for use by the mainstream teacher, as opposed to the computer enthusiast.

How then, has hardware developed during its first decade or so in schools? Perhaps the most obvious change has been in machine memory, which has grown by a factor of a thousand. This has been outstripped by the growth and diversification in storage media. The one to two megabyte (Mb) floppy disc is increasingly used for casual storage while hard discs and optical media approach the gigabyte capacity (1 gigabyte = 1000 Mb). Another diversification has seen microcomputer systems which vary in size from palmtops to networks. Networking, which means connecting computers together so that they can share information, printers, storage devices and so on, was an important development in schools because it allowed IT to become more accessible to pupils (though networks have not been without critics as Chapter 13 discusses). In Science and Technology, a useful alternative has been the stand-alone, mobile computer work-station (Fig. 3.2). This system can be wheeled from room to room, thus helping to solve the problem of access to IT (again discussed in Chapter 13). The system can also include interfacing devices which allow easy connection with laboratory and workshop equipment such as sensors, motors and solenoids.

Other improvements include screen display developments in resolution and colour. Printer output is now often of semi-professional quality, commonly creating possibilities for pupils to produce their own material for public display. Finally, in the area of Science and Technology education there has been a considerable improvement in interfacing equipment used for monitoring and control – this will be the subject of Chapter 12.

Here we have been largely concerned with hardware and its development. This is not because software has been forgotten – the chapters which follow in this section are mainly concerned with educational software in specific categories. One general and important development in software is worth noting here however. This is the growth of the WIMP environment – windows, icons, mouse and pointer. Microcomputers which use the WIMP user interface can be controlled by a mouse rather than a keyboard, removing the need for keyboard skills in some situations.

Future developments in both software and hardware, and their implications for IT use in Science and Technology education will be discussed in Chapter 14. We will now examine in detail the various types of IT use in Science and Technology education. Table 3.1 gives a summary of different uses of IT which are currently prevalent in schools and which have been suggested or discussed in documents from various sources on the place of IT in the curriculum.

Table 3.1 A list of possible types of IT use in Science and Technology education

Databases	Simulations
Spreadsheets	Modelling
Data-logging	CAD
Communications	Word-processing/ desk-top publishing
LOGO	Graphics
Interactive video	Control

Source: Rogers (1990) and Cloke (1988).

CHAPTER 4

CAL/tutorial uses: instructional uses of IT in Science and Technology education

Introduction

We saw in Section A that there are various ways of classifying IT use in education. Probably the most useful framework was that presented by Kemmis *et al.* (1977). This framework was summarised on a kind of continuum in Table 2.1 in Section A and will be used as a framework for the chapters which follow.

'Instructional' uses of IT

Computer programs can be used to teach pupils the structure of the human eye. They can be used to show how light rays pass through different types of lens. One program, entitled ELECTROLYSIS (AVP, Gwent), provides animated diagrams of the migration of ions in different electrolytes such as molten sodium chloride or copper sulphate solution. Students can view the animation or 'freeze' it using the keyboard, and then study simple equations for each animated sequence. In another program, COLOUR MIXING, pupils can examine the results of mixing coloured lights or coloured pigments. Questions can be asked of the student and responses given by the program to their answer. Pupils can study how the internal combustion engine works and learn to identify the different parts of an engine and their main functions using a program called THE FOUR STROKE ENGINE (AVP, Gwent).

All these applications of the computer are *instructional* uses of IT. (This has been traditionally viewed as 'CAL' – see p. 26 – although our definition of CAL, as will be shown, is a much broader one.)

Further details on these, and other examples of instructional CAL, are given at the end of this chapter. Here we examine the nature and types of instructional CAL and discuss some of its strengths and weaknesses.

It is worth distinguishing between two types of 'instructional' uses of IT: drill-and-practice and tutorial use. Both types, as discussed earlier, focus on the subject matter inherent in the program and (as shown in Table 2.1), the locus of control is very much with the software (and hardware) rather than the learner.

'Drill-and-practice'

Consider drill-and-practice first. This mode of learning has a long history associated with words or phrases such as repetition, rote learning, drill, reinforcement and programmed learning. In CAL terms its basic purpose is to provide software which gives practice and repetition in the learning of either processes (skills) or content – most commonly content. This is often done through the use of a highly structured program, taking the user through a certain route, providing reinforcement and feedback along the way.

The drill-and-practice CAL of the 1990s has come a long way from the programmed learning of the 1960s. What advantages can it offer?

Fig. 4.1 A five-year old using CAL

Both learners and teachers who have an ambivalent attitude towards IT can soon become familiar with a drill-and-practice program: it has ease of use. Simply loading the program and following screen instructions is often enough to get started. For many teachers it is a convenient starting point or launching pad to the further use of IT in the classroom. Historically, according to several research reports – see Wellington (1988), for a summary – drill and practice was by far the most common mode of computer use from the early to mid-1980s.

Drill-and-practice CAL can offer a non-threatening, *individualised* learning environment which suits certain learners. Students can proceed at their own pace, will not be subject to verbal rebuke or the body-language equivalent, and can (given the right software) choose their own path through material.

Partly because of the above point, the use of CAL can be *motivating*. Other motivational gimmicks such as smiling faces, cheerful tunes and flashing coloured displays may not be so acceptable and are often seen by teachers as highly distracting in the primary classroom (Ellam and Wellington, 1986). However, the use of rewards and reinforcement, if used carefully and thoughtfully, can increase motivation and attention.

Finally, the use of drill-and-practice can lead to certain basic skills becoming *automatic*. This is a term used by Gagné (1985) who talks of the value of low-level skills, e.g. letter recognition, simple multiplication, becoming automatic. This 'automaticity' is a prerequisite for carrying out higher-level, more complex activities. Through automaticity the learner is able to give more attention to the main tasks in, for example, Science and Technology or to higher-level activities such as problem-solving. Through drill, certain mutual operations become automatic or *routine*. These routines then no longer need the same degree of conscious attention, thus reducing the cognitive load. Thus the acquired, automatic routines can become basic tools in solving higher-level problems.

This summary of a lengthy argument by Gagné is also discussed with great clarity by Sewell (1990). In this context it is worth noting that computer systems themselves can carry out certain 'low-level' tasks in Science and Technology which may then force the learner to move onto a higher plane. This is certainly one of the beliefs in the use of data-logging and spreadsheets, which are discussed later. At the heart of this debate, of course, lies the distinction between authentic and inauthentic labour which has already been introduced (p. 23) and will be considered again. Computer users need to reflect carefully from which tasks the computer is freeing them.

Tutorial programs

A rather more sophisticated version of the instructional use of IT lies in the tutorial program. Again, its basic purpose is to teach a topic by interaction between computer and learner.

With *drill* software, if the learner makes a mistake the program does *not* identify the nature of the error and then go on to provide tuition to remedy it. However, with tutorial programs an increased number of options *will* 'respond' to learners and guide them along appropriate routes through the material.

In a sense, therefore, a good tutorial program (whether simply computer-based or using interactive video) is *responsive and interactive*. As Sewell (1990: 33) summarises it:

> a good tutorial will have an internal model or representation not just of the material to be learned, but also of the likely learning strategies and ways in which learners are likely to respond to the material. In practice, few tutorials for school use have achieved this level of sophistication, although progress is being made in the field of 'intelligent tutoring systems' which aim to satisfy the above requirements.

To our knowledge, few genuine tutorial programs are available for school Science and Technology learning. The software available is largely at the drill-and-practice end of the instructional paradigm. The development of intelligent knowledge-based systems, and artificial intelligence generally, may lead to the introduction of valuable tutorial programs for Science and Technology education, but this is pure speculation. As Sewell (1990) points out, one of the main limitations to the development of genuine Socratic style tutorial programs lies not with the IT itself but with knowledge of learner styles and strategies and the problem of building these into computer-based learning systems.

Criticism and defence of the instructional paradigm

Drill-and-practice CAL is often looked down upon as the inferior use of IT in education. Many people have argued, for example, that to use a 'powerful' computer system for 'mere skill and drill' is rather like using a sports car to go shopping at the supermarket. It has also been argued that instructional CAL encourages the wrong kind of learning e.g. rote learning, the learning of content rather than process – an undue emphasis on factual recall rather than understanding and application. Here we have therefore taken some pains to highlight certain of its advantages, most importantly its value in developing automatic, lower-level skills which are essential in carrying out higher order, more complex activities in Science and Technology education.

It is not our aim to argue for skill-and-drill here but it is worth making one or two further points in its defence. Firstly, there has to be some knowledge base in Science and Technology education (even for strong supporters of the process movement, an issue discussed by Wellington, 1989a). Why not use CAL as an aid in developing a knowledge base, even if only as a complement to the 'chalk and talk' which is still so common in many Science classes? It may well have the novelty value and motivating power which other methods lack.

Fig. 4.2 An example of CAL using a Waves program

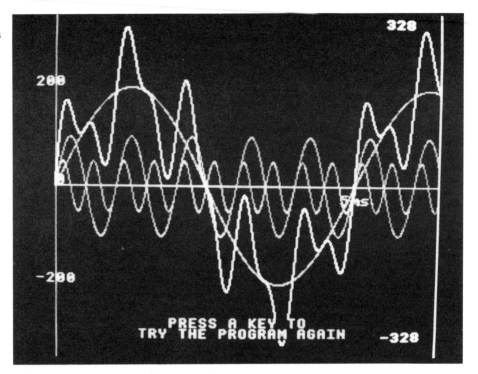

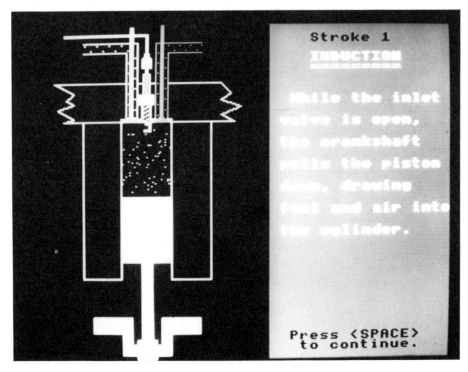

Fig. 4.3 Simulation of a four-stroke cycle

CAL/TUTORIAL USES

Secondly, there is a wide range of instructional CAL available, some of which is quite impressive and may well be educationally quite effective. Not all 'instructional' CAL is concerned solely with the teaching of facts – some of it may be effective in teaching concepts, or perhaps even the skills and processes which are now rightly seen as so important in the Science and Technology curriculum.

It should not be forgotten, however, that instructional uses of computers probably have their roots (psychologically) in the behaviourist tradition of learning and (historically) in the early days of CAL in schools which were influenced by the software and hardware of the time. Educational software had developed little (some argue that this is still the case). The computers of the time, the first to enter schools, were low-powered and slow devices by today's standards.

The uses of IT in Science and Technology education discussed in subsequent chapters are therefore more 'advanced' in terms of:

1. Greater control and initiative by the learner e.g. in a modelling program.
2. More emphasis on the learning *process* (e.g. in creating and using a database), as opposed to *content* (e.g. facts/information from a certain area of Science/Technology).
3. Greater and more varied *interactivity* between the learner and the computer.
4. Encouragement of *collaborative* group work as opposed to individual activity.
5. Encouragement of *investigation and exploration*, e.g. through data-logging, rather than the acquisition of knowledge.

Examples of instructional CAL

A small number of examples of instructional CAL will now be given and which (at the time of writing) are being used in schools (see Table 4.1). The examples are given merely to illustrate the type of skill-and-drill/tutorial software currently available. We are not suggesting that the software below is state of the art or that programs of this kind will always have a place in the Science and Technology curriculum, but the fact that they are being bought for use in schools indicates that teachers are currently finding a place for them.

Table 4.1 Examples of instructional CAL

'Structure and Bonding'
Supplier: AVP, School Hill Centre, Chepstow, Gwent, NP6 5PH
Computers: BBC, Nimbus, PC
'Atomic Structure and Bonding' explains and tests atomic structure, electronic configuration and bonding using a random selection of elements and compounds from the first 20 elements of the Periodic Table. In 'Ionic and Covalent Bonding', the cursor keys are used to select two of the first 20 elements. Providing a simple compound is possible, diagrams are drawn showing its formation. 'Reactivity Series' tests the chemical properties of the most important metals. 'Electrolysis' illustrates the electrolysis of a range of ionic substances, showing how their behaviour is predictable on the basis of the reactivity series.

'The Mole Concept'
Supplier: BBC Software Publications, 80 Wood Lane, London, W12 OTT
Computer: BBC and audio cassette
Using the technique of Computer Synchronised Audio (a voice-track combined with computer software), the concept of a mole is demonstrated and explained. Graphics and animation help to introduce the terms *nucleus*, *proton*, *electron* and *relative atomic mass*. The combining of two hydrogen atoms and one oxygen atom to form one *molecule* of water is illustrated. This leads to the mole as a representation of the number of atoms in 1 gram of carbon-12. The mole is shown to be the vital unit 'ingredient' in the 'recipes' for precisely calculating chemical reactions. This leads to a simulated titration experiment.

Table 4.1 (Cont.)

'Moments'
Supplier: Science Education Software, Dolgellau, Gynedd, LL40 1UU
Computers: BBC, Nimbus
This program illustrates the principle of moments with a simple lever system consisting of a plank of fixed length, a fulcrum and a number of weights. Students may add, move or remove masses to the lever in order to balance it. There are four demonstration options and a problem setting option. Five levels of difficulty are permitted: the more complex ones support practical work with older students. A set of worksheets is supplied with the program to support practical work.

'Electric Circuits'
Supplier: AVP
Computers: BBC (all versions); Arch./A3000; Nimbus
This is a package of four programs: 'Components' gives information concerning electrical components, shows the symbols used in circuit diagrams and provides visual effects. In 'Alternating and Direct Current', the difference between AC and DC is explained and the connection between AC and the sine curve is illustrated. Half-wave and full-wave rectification are introduced, first in terms of the sine curve and then by means of circuit diagrams with diodes. 'Ohms Law' provides circuits with the following arrangements of resistors: single resistance; two resistances in series; two resistances in parallel. Students are able to choose the number of cells and also the value of each resistance. They are then required to calculate the total resistance and the current. Finally 'Circuit Board' provides a simulation of a school circuit board. It allows one, two or three cells, an ammeter, a variable resistance, a switch and any combination of bulbs in up to three parallel circuits.

'Atomic Theory'
Supplier: AVP
Computers: BBC, Archimedes, Nimbus
This is a suite of seven games to complement other work on Atomic Theory. Topics covered include electron orbits, an atomic fruit machine game, the nucleus, elements, symbols and radicals, ionic bonding and naming compounds.

'Balancing Equations'
Supplier: Longman Micro Software, Layerthorpe, York, Y03 7XQ
Computers: BBC, Nimbus
This is a revision and practice program in which students have the opportunity to supply correct formulae for compounds in a chemical equation and to balance the equation. It is possible to balance equations for which correct formulae have already been supplied.

'Uniformly Accelerated Motion'
Supplier: BBC Software Publications
Computer: BBC
This program employs the tutorial technique of Computer Synchronised Audio (voice-track used in combination with the computer). It can be used for direct learning or for revision. The essential theory of the subject is explained and illustrated not only by graphs but also by graphics animations. The program assumes the most elementary knowledge on the part of the user, but gives a thorough treatment of uniform velocity, uniform acceleration, the idea of terminal velocity and the acceleration due to gravity. The fundamental equations of uniform velocity and uniformly accelerated motion are explained and formulated. A number of problems based on uniformly accelerated motion are included.

CAL/TUTORIAL USES

Table 4.1 (Cont.)

'The Human Skeleton'
Supplier: AVP
Computers: BBC, Archimedes
This is a set of diagrams. Users can look at the skeleton from front, side and rear and also at several individual components, including the skull. This program gives a set of views of the skeleton, with diagrams of the skull, vertebrae, ribs, girdles and limbs.

'The Four-Stroke Engine'
Supplier: AVP
Computers: BBC, Nimbus
The main parts of the engine are identified, with a description of their functions. The four-stroke cycle is demonstrated in stages for a single cylinder, including the valve action, the states of the gases and the crank positions at each stage. An animated simulation of the cycle can be selected. There is a tutorial section after the demonstrations. The program illustrates two advantages over a traditional approach to teaching about the four-stroke cycle using posters; these are the animation facility and the option of referring back to the demonstrations as needed during a user's attempts at the tutorial questions.

'Gravity Pack'
Supplier: Cambridge Micro Software, CUP, Shaftesbury Road, Cambridge, CB2 2RU
This package contains three programs that simulate the motion of objects under the influence of gravity. The programs allow more control than is possible in laboratory experiments, so users can explore the characteristics of gravity and gravitational fields more effectively. 'The Monkey and the Hunter' is an interactive animation of a hunter trying to shoot a monkey and shows how bodies of different masses are accelerated at the same rate by gravity. 'Newton's Cannon' demonstrates that if a cannonball is fired horizontally from a height it will either crash to Earth, shoot off into space or go into a stable orbit, depending on the velocity at which it is fired. 'Satellites' explores the behaviour of a satellite in a gravitational field between the Earth and the Moon. (This program is more suitable for older students.)

'The World of Newton'
Supplier: Longman Micro Software
This computer simulation allows the user to investigate a microworld unfettered by the complications of the real world. The program uses a small oblong object which can be moved around the screen with 'kicks'. The booklet tells you that this object 'moves on the screen according to Newton's Laws of Motion'. Different parts of the program provide entertaining challenges, e.g. 'Crazy Maze', 'Beat the Clock'. Ideally, a networked system of computers would make it possible for several groups of pupils to be using the program simultaneously. The simulation illustrates the capacity of computer to model idealised situations. Frictionless motion, for example, can only be approximated in a laboratory but it can be simulated with ease on a computer screen; there are several computer and video games which use this same idealised model.

'Predator-Prey Relationships'
Supplier: AVP
How do the populations of predators and prey change over yearly periods? This program lets users investigate. The user supplies numbers, size, population, etc. The program predicts predator–prey relations over annual cycles.

'Population Pack'
Supplier: Longman Micro Software
The pack contains programs on human population growth and on a Malthusian model of population, food and energy supplies.

Table 4.1 (Cont.)

Three programs on genetics and evolution

1 **'The Blind Watchmaker'**
   Supplier: SPA, PO Box 59, Leamington Spa, Warwicks

2 **'Insect'** from Beebug (also *Beebug Magazine*, Vol. 6, Issue 1, May 1987)

3 **'Survival of the Fittest'**
   Supplier: SPA Program (3) investigates genetic principles underlying evolution. It includes a population division facility. All these programs offer an insight into the process of evolution which would otherwise be extremely difficult to illustrate.

Programs (1) and (2) demonstrate that variations in organisms are due to reproduction in which gene mutation occurs. They can be fascinating for almost any age! The user plays the part of the environment, determining which organisms are adaptively favoured.

CHAPTER 5

Simulations

Introduction

The second paradigm, revelatory, concerns the use of simulations, through which reality is gradually 'revealed' to the learner. This chapter discusses the use of simulations in learning Science and Technology.

Types of simulation in Science and Technology education

It is useful to make some fairly crude distinctions between types of simulation. These types are debatable but should act as a rough guide:

1 Direct copies of existing laboratory activities, e.g. titrations.
2 Simulations of industrial processes, e.g. the manufacture of sulphuric acid, bridge building.
3 Simulations of processes either:
   - too dangerous
   - too slow, e.g. evolution, population growth, an ecosystem of any kind
   - too fast, e.g collisions
   - too small, e.g. sub-atomic changes to be carried out in a school or college environment
4 Simulations involving non-existent entities, e.g. ideal gases, frictionless surfaces, perfectly elastic objects.
5 Simulation of models or theories, e.g. kinetic theory, the wave model of light.

This does not form an exhaustive list, but at least provides a framework for looking at the advantages of simulations. Examples of each of these types of simulation are given at the end of the chapter.

Advantages of computer simulations in Science and Technology education

The main advantages of using simulations can be summarised as follows:

1 *Cost*: Money can be saved in directly copying some laboratory experiments, either by reducing outlay on consumables (e.g. chemicals, test-tubes), or by removing the need to buy increasingly costly equipment in the first place.
2 *Time*: Using a computer simulation instead of a genuine practical activity may save time, although some teachers are finding that a good computer simulation in which pupils fully explore all the possibilities may take a great deal longer.
3 *Safety*: Some activities simply cannot be carried out in a school setting because they are unsafe.
4 *Motivation*: There is a feeling, though with little evidence to support it, that computer simulations motivate pupils in Science and Technology education more than traditional practical work.
5 *Control*: The use of a simulation allows ease of control of variables, which traditional school practical work does not. This may lead to unguided discovery learning by pupils who are

encouraged to explore and hypothesise for themselves (i.e. Kemmis's revelatory paradigm).

6 *Management*: Last, but certainly not least, computer simulations offer far fewer management problems to teachers than do many traditional activities. Problems of handing out equipment, collecting it back again, guarding against damage and theft are removed at a stroke. Problems of supervision, timing and clearing up virtually disappear. Indeed who needs an expensive and noisy laboratory, it might be argued?

Dangers of simulation

So much for the supposed advantages of computer simulations. What of the dangers in using computer simulations in Science and Technology education? The main dangers of using simulations lie, in our view, in the hidden messages they convey. We have classified them as follows:

1 *Variables*: Simulations give pupils the impression that variables in a physical process can be easily, equally, and independently controlled. This message is conveyed by simulations of industrial processes, ecological systems and laboratory experiments. In reality not all variables in a physical situation can be as easily, equally, and as independently controlled as certain simulations suggest.

2 *Unquestioned models, facts and assumptions*: Every simulation is based on a certain model of reality. Users are only able to manipulate factors and variables *within* that model. They cannot tamper with that model itself. Moreover, they are neither encouraged nor able to question its validity. The model is hidden from the user. All simulations are based on certain assumptions. These are often embedded in the model itself. What are these assumptions? Are they ever revealed to the user? All simulations rely on certain facts, or data. Where do these facts come from? What *sources* have been used?

3 *Caricatures of reality*: Any model is an idealisation of reality, i.e. it ignores certain features in order to concentrate on others. Some idealisations are worse than others. In some cases, a model may be used of a process not fully understood. Other models may be deceptive, misleading or downright inaccurate; they provide caricatures of reality, rather than representations of it.

4 *Confusion with reality*: Pupils are almost certain to confound the programmer's model of reality with reality itself – such is the current power and potency of the computer, at least until its novelty as a learning aid wears off. Students may then be fooled into thinking that because they can use and understand a model of reality they can also understand the more complex real phenomena it represents or idealises. Perhaps more dangerously, the 'microworld' of the computer creates a reality of its own. The world of the micro, the keyboard and the VDU can assume its own reality in the mind of the user – a reality far more alluring and manageable than the complicated and messy world outside. The 'scientific world' presented in computer simulations may become as attractive and addictive as the microworlds of arcade games as noted by Weizenbaum (1984), and Turkle (1984).

5 *Double idealisations*: All the dangers and hidden messages discussed so far become increasingly important in a simulation which uses a computer model of a scientific model or scientific theory which itself is an idealisation of reality, i.e. the idealisation involved in modelling is doubly dangerous in simulations which involve a model of a model. A simulation of kinetic theory, for example, is itself based on a model of reality.

Summary

The above discussion has highlighted some of the advantages and dangers in using simulations in Science and Technology education. More could be added, and more will be revealed as the use of computer simulation increases. Given that Science teachers will continue to use them, what safe-

SIMULATIONS

guards can be taken to reduce the messages of the medium?

Firstly, all teachers and thereby pupils, must be fully conscious that the models they use in a computer simulation are personal, simplified and perhaps value-laden idealisations of reality. Models are artifacts. Students must be taught to examine and question these models.

Secondly, the facts, data, assumptions, and even the model itself which are used by the programmer must be made clear and available to the user. This can be done through a teachers' guide, or the documentation accompanying the program. All sources of data should be stated and clearly referenced. Any student using a simulation can then be taught to examine and question the facts, assumptions and models underlying it.

By building these safeguards into the use of Science simulations, this type of CAL in Science may then move slightly towards the right of the continuum shown in Table 2.1 (p. 26). Learners will be more in control of their own learning, rather than being controlled by the computer, or rather the computer programmer. The use of modelling in Science and Technology education, discussed shortly, is a classic case.

Examples of simulation programs

Here is a selection of examples of simulation programs for use in Science and Technology education (Table 5.1). A wide range of simulations is now available for school use, ranging from simulations of chemical collisions, the manufacture of ethanol or the siting of a blast furnace to the simulation of electric and magnetic fields, electricity use in the home, wave motion, floating and sinking, a 'Newtonian' world of frictionless movement and the construction of bridges. For the Life Sciences, simulations are available on pond life, the human eye, nerves, the life of the golden eagle and predator–prey relationships.

Table 5.1 Examples of simulation programs

'Bridge Building'
Supplier: Longman Micro Software
Computer: BBC
The program allows students to evaluate three different types of bridge (beam, Warren, arch). Pupils can choose different features of the bridge (e.g. material, arch depth etc.), and evaluate the different factors important in bridge building.

'Make a Million'
Supplier: AVP
This is an educational game involving the industrial production of a variety of substances and electricity from basic raw materials. The processes involved are blasting, electrolysis, fractionating, generating electricity and making and adding acid. The program is a game in which the user chooses his pathway to financial success with coloured graphics and sound throughout. A set of worksheets containing information about the processes covered in the program is provided with the package.

'Watts in your Home'
Supplier: Cambridge Micro Software
Computers: BBC (all models)
This is a simulation program showing energy/electricity use in a typical home. It provides an interesting way of comparing the costs of energy-consuming appliances in domestic use.

Table 5.1 (Cont.)

'Molecules'
Supplier: Cambridge Micro Software
This is a simulation program allowing pupils to experiment with particles inside a vessel. The simulation is based on the kinetic theory model of matter and allows a number of variables to be altered (e.g. pressure, volume, temperature), though not all at once! Students should be told that this is a computer *simulation*, based on a model.

'Three ecological simulations'

1 **'Ecology Foodwebs'**
Supplier: AVP
This is a three stage package: an animated guide to foodwebs, an interactive section and a quiz. It uses games style presentation.

2 **'Lake Web' and 'Bio-Wood'**
Supplier: AVP
These are similar programs which challenge the user to survive in an environment. Survival depends on deciding what to eat, what to ignore and what to escape from.

3 **'Golden Eagle'**
Supplier: AVP
The user runs a wildlife reserve with the aim of doubling the golden eagle population. It encourages systematic record keeping and learning from repeated attempts.

'Waves'
Supplier: Resource, Exeter Road, Wheatley, Doncaster
Computer: BBC (available on $5\frac{1}{2}$ in. discs)
'Waves' *1* and *2* (written by R.A. Sparkes) is a very useful set of simulations, especially for reinforcing basic points which teachers hope to show in practical work. With little initial guidance, pupils can explore the behaviour of single pulses in a range of situations. Waves can then be investigated by changing from single pulses to sets of pulses. Both longitudinal and transverse pulses and waves are possible. Topics which can be investigated include reflection, interference, refraction, ripple tanks, Young's fringes and Lissajous figures. This set of simulations illustrates the computer's potential for graphical analysis. The complex and rapid changes in wave motion can be slowed down and analysed as a series of simple graphical elements in the form of pulses.

'Motion in Space'
Supplier: Cambridge Micro Software
Computer: BBC B or Master
This program simulates space walking and space missions in which a *female* astronaut can be moved around by firing four gas jets. As the astronaut moves, the horizontal and vertical speeds are given allowing the components of motion to be studied in some detail. The fuel used and oxygen supply remaining are given on the screen. Having learnt to space walk, an interesting experience in itself, the user can then try a space mission. Three of the missions involve space rescues, two involve rebuilding a damaged satellite. Motion can be studied in the idealised world of frictionless movement and perfectly elastic collisions. It is enjoyable and, if used carefully by the teacher, educational.

SIMULATIONS

Table 5.1 (Cont.)

'The Ethanol Question'
Supplier: BP Educational Service, Blacknest, Alton, Hampshire, GU34 4BR
Computers: BBC, Nimbus
This is a simulation designed to introduce the difficult concepts of rate of reaction and equilibrium in an industrial context where chemical reactions seldom proceed to completion. The simulation is in two parts. Part 1 guides students through the different factors which influence the position of equilibrium in the reaction so that they can make the most of the raw materials. Part 2 introduces the costs associated with changing conditions and the task of producing ethanol at a competitive selling price, with guidelines given for a target price. The package includes a students' booklet, worksheets and a guide for teachers.

CHAPTER 6

Modelling

Introduction

What is computer modelling? The word 'modelling' is used here to mean the representation of systems and processes by a computer. The main ideas in modelling come from the user; this contrasts with computer simulations, in which many of the ideas are supplied by the program author. Some people see modelling as a useful process in itself; this view might be held by some Mathematics teachers, for instance. Alternatively, modelling might be seen principally as an approach to problem solving – a view which is more within the realms of Science and Technology education.

Modelling in problem solving

Here is an exercise in problem-solving: a water tap has been left running – how much water will come out? This problem could be used in Science and Technology education to focus on any of the following:

- practise measuring quantities
- handle corresponding sets of data (time and quantity of water)
- correlate data
- design and carry out a plan for solving the problem
- make predictions about the actual situation
- make general predictions
- criticise the approach used to solve the problem

Even an apparently simple problem like this is rich in educational potential including possibilities for using computers. It is instructive to consider in which of the activities listed above could computers usefully be used. When children first begin, informally, to correlate observations they go through a heuristic modelling process. The model might start in a form such as this:

1 There is more water now than there was before.

The teacher would encourage children to make more quantitative observations, which in time might lead towards statements like these:

2 It took 2 minutes to fill the bucket.
3 We got 1 full bucket every 2 minutes.

By now an *algorithmic model* is emerging; this contains a reliable rule which makes a connection between time and quantity of water. Such an algorithm could be used for predicting:

4 It would take 20 minutes to get 10 buckets of water.

The algorithm used in statement (4) could be expressed as follows:

5 Double the number of buckets to work out how long it takes for them to fill.

It could also be expressed in algebraic form:

6 Time = 2 × buckets.

Statement (6) represents the algorithm in a form

which could be used with a computer. It is a *computer model* of the water-flow problem.

The purpose of the description above is to show how a simple computer model is actually part of a modelling process in which children are already engaged. What distinguishes statement (6) from earlier statements is its *language* – algebraic in this case. There is no greater *meaning* in statement (6) than in statement (3).

Computers can respond to operational statements, such as statement (6), rather than to conversational language.

Any system, event or process which can be described in terms of arithmetical, algebraic, differential or statistical operations, can be modelled. For Science and Technology teachers, this is both an opportunity and a challenge. A great many relationships exist in school Science and Technology which can be modelled – but can pupils understand these scientific relationships when they are expressed in the form needed by computers? For example, which of statements (3) and (6) would children find easier to understand?

The challenge for the teacher is to develop pupils' understanding and confidence in a scientific model and then to help them across the language barrier to the computer model.

Modelling applications

Good modelling software or applications will probably be free of content. This allows the user total freedom of choice about the nature of the model he or she creates. What, then, does the application do? The application should make the user-interface friendly and helpful. Writing the model should be easy. Manipulation of the model (for example, editing) should be straightforward. And the output-interface should be simple and clear. The application thus transfers the 'inauthentic labour' of programming, calculating and presentation to the computer, leaving the user to focus on the intellectually creative task of devising and exploring the model.

Types of modelling applications

Because of their flexibility and ease of control, modelling applications are often used to examine how systems change from some known state to some future state. The model can be refined and altered to find out what changes occur.

Some applications focus on the system's *process of change* or evolution. Spreadsheet models are examples of this; using a spreadsheet or cellular model, components of the model can be observed, and intermediate states can be investigated. For example, in a spreadsheet model of a predator–prey relationship it would be possible to follow simultaneously the changing population of adults and offspring of both prey and predators. One particular spreadsheet modelling application, the CELLULAR MODELLING SYSTEM (Ogborn, 1986), is illustrated in Figure 6.1. The screen displays a set of cells containing numbers, variables, formulae or graphs. In the example, Nucleus A decays, forming Nucleus B, which in turn decays to form the stable Nucleus C. The initial sample was set with A at 1000 nuclei and B and C at zero. The application repeatedly works through the set of cells, re-calculating values from the formulae. The figure displays the result of 100 such loops.

Other applications focus on the outcome of the model. The computer display might use a split screen to show the model as a set of equations alongside a graph or chart of the outcome. An example is the discharge of a capacitor, where interest is often centred on the graph of potential difference against time. The two examples below use a modelling program called the DYNAMIC MODELLING SYSTEM (Ogborn, 1984). This program was originally integrated into the Nuffield A level Physics course. Its use has subsequently spread into many other courses.

Figure 6.2 shows the computer screen split into halves with the model on one side and the graphical output on the other side. The system which has been modelled is an electric circuit consisting of a resistor, an inductor and a charged capacitor.

Figure 6.3 also shows the model on one side of the screen and a graphical display on the other half. This example contains two distinct parts,

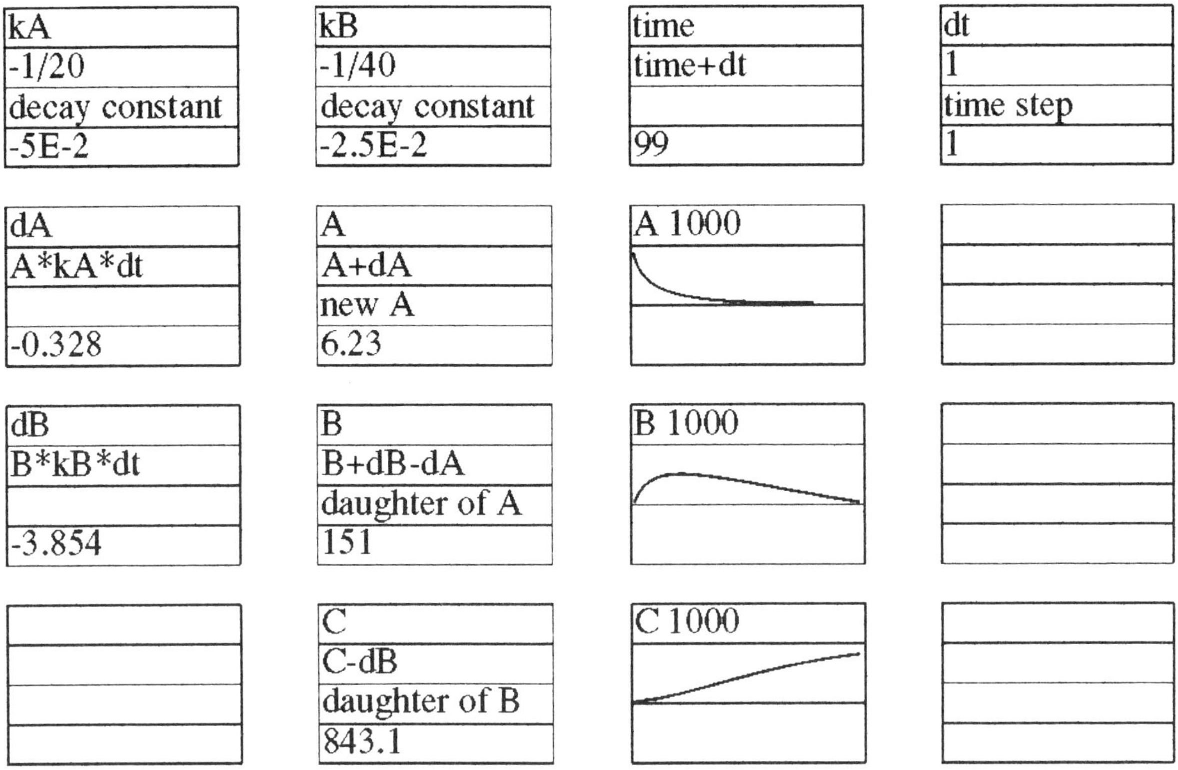

MODELLING

kA	kB	time	dt
-1/20	-1/40	time+dt	1
decay constant	decay constant		time step
-5E-2	-2.5E-2	99	1

dA	A	A 1000	
A*kA*dt	A+dA		
	new A		
-0.328	6.23		

dB	B	B 1000	
B*kB*dt	B+dB-dA		
	daughter of A		
-3.854	151		

	C	C 1000	
	C-dB		
	daughter of B		
	843.1		

Fig. 6.1 CELLULAR MODELLING SYSTEM. Model of radioactive decay chain

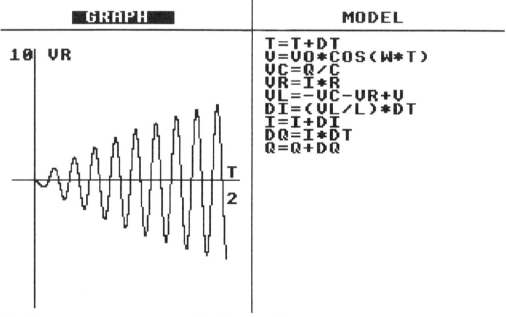

Fig. 6.2 DYNAMIC MODELLING SYSTEM. Oscillator model

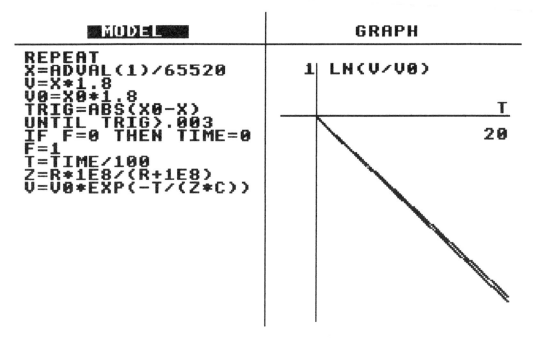

Fig. 6.3 DYNAMIC MODELLING SYSTEM. Comparison of model with laboratory data

firstly taking a series of measurements in the laboratory and then, in the last three lines of the model, overlaying the experimental data with the results from an exponential equation. This program can accommodate the plotting of functions as well as variables; in this case a logarithmic function was chosen because it highlights the extent of agreement between measured and calculated data.

Selection of the appropriate application is a matter for the teacher's judgment; for instance if we were addressing the question of how a mass moves on the end of a vertical spring we would probably prefer the 'outcome' oriented application; on the other hand the spreadsheet approach would better suit questions about how the amounts of yeast, sugar and alcohol vary during fermentation.

Another type of modelling application focuses on the *operations* contained in the model. The use of the language LOGO to control the movement of a screen-based or floor-based object is an example. The emphasis here is on understanding how to use operations to achieve a target outcome. Applications in technology include control and robotics. For a more general description of modelling in Science education see Ogborn (1990).

The three examples above are pitched at a fairly high level in school Science; they would probably occur in the post-16 curriculum. Teachers may worry, initially, that they are too advanced. In our experience, however, the unfamiliar medium of the modelling applications turns out to be less of a challenge to pupils than to teachers. We have found Science students who have been relatively mathematically inexperienced nevertheless able to cope with modelling second-order differential systems such as oscillators. The differential equations are re-written in the model in the form of linear difference equations, which have the advantage of being mathematically simple and plausible. An example could be the relation between velocity, displacement and time, which is formally written as $V = ds/dt$ but which appears in the model as $DS = V \times DT$. Less numerate students have not been put off by this, especially

MODELLING

when they talk it through in the form: 'a little bit of displacement equals velocity times a little bit of time'.

Needless to say, modelling applications are also well received by mathematically competent students, to the extent that once guided through an introductory model, they can become confident, independent and creative in using the software.

CHAPTER 7

Databases

Introduction: what is a database?

This chapter will discuss the uses of databases in Science and Technology education by considering what a database is and what it can offer.

In its simplest form a database is nothing more than an organised collection of information. Thus an address book, a telephone directory, a card index, and a school register are all examples of databases. They all contain data, which can convey information to people, and which is organised in a more-or-less systematic way, e.g. in alphabetical order. The advantage of organising data is partly for ease of use and access to information, but also depends on the fact that well-organised and structured data can be used to show patterns and trends and to allow people to make and test hypotheses or hunches. Therein lies the *educational value* of a database. Having an organised and clearly structured collection of data allows and even encourages people to derive information and knowledge from it. Examples will be given later.

This chapter will be concerned primarily with databases which are stored and presented using a computer system but it should be remembered that the discussion could often relate equally to databases stored and presented using only ink and paper, which possess at least partly as much educational value. Indeed the creation and use of non-computer-related databases can often be a valuable and complementary activity in relation to computer work.

The advantages of storing, organising and retrieving information from a computer system are worth considering briefly. Firstly, using magnetic or optical media (e.g. floppy discs, 'laser discs', CD-ROM etc.), huge amounts of data can be stored in a relatively compact form. Secondly, data can be retrieved from a computer database quickly. Thirdly, data retrieval from a computer database is relatively flexible. For example, to find a number in a paper-based telephone directory from a name and initials would be almost as quick as finding it from a computer-based directory. But consider the situation in reverse. How long would it take to find a person's name and initials with only their number? With a suitable computer database this could be done as quickly as a search in the other direction. Fourthly, changes (editions, additions and subtractions) to a computer database can be made more easily and more painlessly than to, say, a card- or paper-based database – in a way similar to the use of word-processors in amending and re-drafting text.

The huge storage potential, speed, and flexibility of computer databases all have implications for their educational value which is exemplified later.

Records, fields and files

Certain terms are generally used in connection with databases and these are worth considering briefly. A *file* is a collection of information on one topic (e.g. dinosaurs). This file might be organised into separate *records* (e.g. each type of dinosaur with its own name). Within each record data might be stored on each kind of dinosaur and this can be organised into *fields*. One field might contain data on what the animal eats, another on its size, another on its weight, and so on. In setting up a file as part of a database, people can decide how many records they wish to include (e.g. how many different dinosaurs, and how many fields they wish to use in storing information on each animal). Of course, they can always add records (e.g. if we hear about more dinosaurs, or more fields or if we decide to store new or more complex data). Thus records and fields can be added to, edited, or even removed.

Databases in Science and Technology education

What can databases be used for in the Science and Technology curriculum? A general framework would include the following:

1 *For recording data collected during an investigation or an experiment.* Data can be entered directly onto the database and stored on a computer-based medium (e.g. a disc).
2 *In allowing students to sift or browse through their own, or someone else's data using the computer.* This kind of serendipitous learning (learning by browsing) can often be very valuable and is commonly underestimated.
3 *Students can explore data in a more systematic manner.* They can:

- look for patterns
- put forward hunches
- make predictions
- suggest and test hypotheses
- draw and discuss interpretations

4 *Better display.* With suitable software the computer system can be used to display and present data so that it conveys information in an attractive and clear way (cf. spreadsheets, discussed in Chapter 11).

The use of databases in Science supports and enhances many of the so-called process skills in the Science curriculum such as classifying, hypothesising and testing. This applies equally to textual and numerical data stored on the database. To make these general points more vivid, specific examples of databases will be given below.

Some interesting classroom research has been carried out on the use of databases in education, although little on their use in Science and Technology education. Two examples are summarised briefly below. Smart (1988), in an article nicely entitled 'The database as a catalyst', reports on a case-study of the use of QUEST, a database program written for school use, with middle-years children. He claims that the use of a database:

- encouraged exploratory talk in small groups around the keyboard or print-out
- gave more responsibility to the learner
- developed skills in hypothesis forming and testing

Smart suggests that the computer database removed some of the drudgery of testing, then modifying and re-testing hypotheses – perhaps this is the inauthentic labour referred to on p. 23.

Another small-scale study of database use (Spavold, 1989) comments on the need for peer-group support in using databases in the classroom and also the inhibiting effect of the lack of keyboard skills. These are both important points for classroom teachers to bear in mind. Spavold also comments on the value of children constructing and compiling their own database in addition to just interrogating one – the compilation process assists in understanding field and record structures.

Database programs and ideas for their use

There is a wide range of databases available for use in schools at all levels. Certain favourites become apparent, and in schools in our own locality the program GRASS (Newman College, Birmingham) was commonly used at primary, middle school and secondary level. Incidentally this may not mean that it is the *best* database – it simply means that it has become widely used

DATABASES

perhaps for other reasons (e.g. availability, dissemination through INSET etc.).

Other databases in UK schools include QUEST, KEY, FIND, DIY, OURFACTS and EXCEL. Our intention in the examples below is not to give an exhaustive description of available databases but to illustrate uses, with examples of datafiles (Table 7.1). Included are a few suggestions for pupil activities which we have written in a style appropriate for pupils. The examples section ends with a list of possible database programs which pupils can use to construct their own database, and a suggestion for pupils' instructions on creating a database on electrical appliances.

Table 7.1 Examples of databases

1 **'Key'**
Supplier: Anglia TV, Anglia House, Norwich, NR1 3JG
Computers: BBC, Key Plus: Archimedes, Nimbus
'Key' is a data handling package which is ideal for school pupils, and can be used with the following datafiles: Energy, Our Neighbours in Space, Rocks and Minerals, Acid Drops, Birds of Britain, Mammals, Materials, Minibeasts, Periodic Table, Weather and Climate, World Population.

The datafile 'Life and Death of a River' allows users to explore a case of river pollution and to see what can be done to clean it up. 'Weather and Climate' is a datafile with data from 80 weather stations around the world. Pupils use this to analyse world weather patterns. Diet and fitness can be assessed using the datafile 'Fit to Eat'. Users can analyse the food they eat and estimate fitness in terms of flexibility, strength, endurance and body fat.

2 Other databases on food
Pupils can also investigate and analyse the nutritional content of foods with the databases 'Micro-diet' available for BBC micros, from Longman Logotron, 'Food' for BBC micros, from Hutchinson Software, 'Balance Your Diet' from Cambridge Micro Software or 'Diets' for BBC micros, from AVP.

3 **'Electricity in the UK'**
'Interactive database on the use of energy and electricity in the United Kingdom'
Supplier: UKAEA Education Service, Building 354 West, Harwell Laboratory, Oxfordshire, OX11 0RA
Computers: BBC, Nimbus
This database contains detailed information about the past and present use of energy and electricity in the UK. It also acts as a 'simulated' laboratory that allows users to estimate the likely future demand for these commodities and see how these demands might be met.

The basic data are capable of being interpreted, analysed and extrapolated at a range of levels. It is accompanied by a Teachers' Guide.

4 **'BP Energy File'**
Supplier: BP Educational Services
Computers: BBC, Nimbus
'BP Energy File' is a database of world energy use produced by BP Educational Resources. It is regularly updated using data from readily available sources.

Pupils can use this program to study energy use in different places and of different fuels.

5 **'Chemdata' and 'Periodic Properties'**
'Chemdata' (Longman Micro Software, York) and 'Periodic Properties' (Hodder and Stoughton, Sevenoaks) programs both allow students to 'discover' the relationships between the properties of the chemical elements. Both can display data in a table or in graphical form. Datafiles are provided on melting point, boiling point,

Table 7.1 (Cont.)

density and many other properties. 'Chemdata' should be used to explore the properties of different elements. 'Periodic Properties' allows exploration of the different properties of the elements in the Periodic Table. Patterns in the properties of the elements are apparent. Predictions can be made about some of the elements' properties.

6 **'Earth in Space'** (database and teletype simulation)
Supplier: Resource
Computer: BBC
'The Earth in Space' curriculum package includes two discs. One is a teletype simulation, which simulates information and beliefs related to the solar system (e.g. Galileo's views compared with those of the Church). The other disc stores a database (DIY base) of facts on the solar system. The package contains a large number of activities and teachers' notes related to the two discs. It covers the following areas: historical ideas; the night sky; planetary motion; the Sun and planets; the Universe.

'Science, Religion and the Solar System' can be used to try a 'teletype simulation'. This gives 'news' on the computer screen as if it had just happened (e.g. Ancient Chinese beliefs about the sky dragon, Egyptian explanations of why the Moon waxes and wanes and Copernicus' book of 1543). The teletype messages on the screen can be used to read about Galileo's disagreement with the Vatican. It answers the questions: when did Newton publish his famous book?; what was it called?; which side did it support – Galileo or the Vatican?

'The Planets' is a database in the 'Earth in Space' package to explore the different planets. Start from the menu called 'Look at Planets'.

'Is our Planet Unique?' (DIY base) can be used to find out some of the main points about the Earth, how it differs from other planets and if it has anything in common with another planet. Notes can be made under each heading.

7 Pupils constructing their own database
Any database program can be used here, for example FIND, DIY, KEY and GRASS, or any other database program which a school can acquire. Pupils can, for example, make their own database of domestic electrical appliances. For each appliance, notes can be made of the name, its power rating (in watts) and the average time it is used.

The database can be used to compare different appliances.

CHAPTER 8

Data-logging

Introduction: what is computer data-logging?

By the word 'logging' we mean the recording of information. Data-logging implies that the process is systematic and the information is structured. Computer data-logging has so much to offer to Science and Technology teachers that, on first glance, it seems inexplicable that its use has hitherto been far from widespread. In this chapter we will consider why this has been the case, and whether the situation is likely to change. We will also ask what data-logging actually is, and how it has been – and might be – carried out in schools.

Scientists and Science students have logged data by hand for generations. Data-logging by computer has been around in research and industry for two decades; it first occurred in schools in the UK in the early 1980s.

What can a computer add?

The literature on IT in education, from learned journals to manufacturers' brochures, frequently stresses a host of positive reasons why a particular process or product should be used in schools. There are some occasions where it seems to us that the case is merely an elaborated version of the argument 'because it's there'. This could be fairly said about the first few years of data-logging in schools, when many teachers' answers to the above question would have been 'nothing!' or 'unwanted hassle!' There is, however, broad agreement about what computers *should* add to the process of data-logging. The following advantages have been claimed:

- *Speed.* Computers can often log much faster and more frequently than humans.
- *Memory.* Computers have enormous capacity for retaining and accessing a large body of data in a compact form, for instance on a magnetic disc.
- *Perseverance.* Computers can keep on logging – they do not need to stop for sustenance or sleep.
- *Manipulation.* The form in which data are gathered may not be the form in which we want to communicate. Computers come into their own when it comes to fast manipulation of large bodies of data.
- *Communication of meaning.* Computers can present data when gathered, in realtime, using graphic display to enhance the meaning which is communicated to the observer.
- *Stimulating.* Students' imaginations and curiosities are likely to be stimulated by the use of computers. In other words, there is a message in the medium of IT which teachers might use positively to enhance learning.

Some of these advantages are aimed at transferring 'inauthentic labour' from the human to the machine. It has been argued, for instance by Barton and Rogers (1991), that the change of emphasis away from the routine process of logging towards the use of interpreting skills enhances scientific thinking, creativity and problem-solving ability. This view is not universally shared by teachers. It has been pointed out to us that perseverance, ability to organise data systematically and calculating skills are part of Science and Technology and that students should go through these processes in practical work. As a Chemistry teacher put it: 'we want students to manually collect data and plot results'. This might be an uncritical attachment to outmoded tradition – or a recognition of some key core activities which are under threat from the introduction of IT into practical work.

Teachers and data-logging – future allies?

Data-logging should have established a place in the curriculum in the UK in the mid-1980s, with the appearance at a low price of a VELA (Versatile Electronic Laboratory Aid) (Fig. 8.1). For many schools however, that was both the beginning and the end of data-logging, for in spite of VELA's impressive specifications, many of its owners had problems with it. VELA was ahead of its time; its audience needed help and guidance, not channels and ports. It has taken schools several years to recover from this misguided introduction to data-logging.

Until very recently, what has been missing as a design consideration in data-logging equipment is what is arguably the most important element – user-friendliness. Hardware has been designed by microelectronics engineers who are experts in their field but who poorly appreciate the classroom, school laboratory and workshop situation. Teachers seem to have been left to pick up the pieces. In our awareness, some teachers have done this remarkably well, with little support and a lot of enthusiasm. For the majority, however, the payoff has not justified the perceived labour needed to get to grips with an unfriendly technology.

If data-logging in schools has had an inauspicious past, what of the future? Perhaps surprisingly, future prospects look very much better. There are two reasons for this:

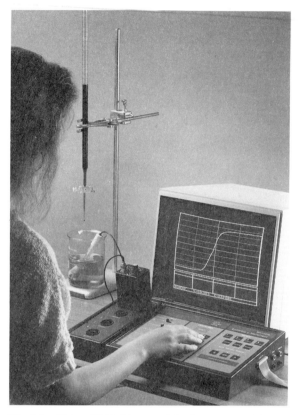

Fig. 8.1 Using 'VELA', one of the longest established devices for school laboratory monitoring

1 Data-logging has been formally allocated a role in the curriculum in the 1990s.
2 Hardware and software designers have begun to recognise that if IT is to be successful, it must be friendly for the non-specialist user.

Data-logging design

Perhaps the simplest data-logging system is

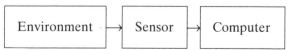

The sensor plays the part of a translator. It responds to some property of the environment and sends a message to the computer. The message,

DATA-LOGGING

or signal, has the form of a voltage at one of the computer's input ports. The computer is programmed to record the value of the input signal. Temperature is an example of an environmental property which can be sensed in this way. There are many others, as a glance through the Science and Technology suppliers' catalogues will reveal.

The simple system on the previous page begs many questions!

- How does the sensor connect to the computer?
- How does the computer collect, store and present data?
- Does the sensor's output signal match the computer's input port?
- Does the sensor need power?
- What is being sensed?
- How fast and for how long are data to be logged?
- How much does it all cost?

The list, which is virtually endless, contains questions which most Science and Technology teachers and students do not want to have to answer – they simply want to use the computer! Consequently, recently designed data-logging systems have incorporated their own answers to these questions. Sensors can identify themselves, logging rates are automatically optimised, and interfaces match the type of information given by the sensor to the type which the computer can accept. An improved system is shown opposite.

Teachers should expect many of these features in new data-logging equipment. The result should be that the inauthentic labour of matching the computer to the environment is removed from the teacher and is incorporated in the hardware and software design of the system.

Data-logging in schools

We do not intend to give an exhaustive survey of data-logging practice in schools. Instead we will highlight some cases, perhaps because they have already earned a place in the curriculum, or because they are innovative and exciting. An appendix, necessarily incomplete, is included for further sources of information (see p. 68).

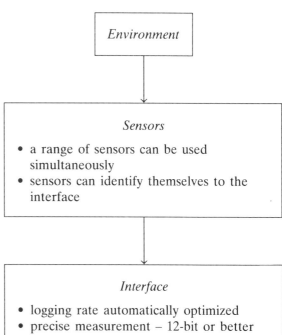

Environment

Sensors
- a range of sensors can be used simultaneously
- sensors can identify themselves to the interface

Interface
- logging rate automatically optimized
- precise measurement – 12-bit or better sampling
- long-life lightweight re-usable power source
- remote operation – computer not needed for logging
- signal to computer is 'conditioned', including calibration, matching and buffering for safety

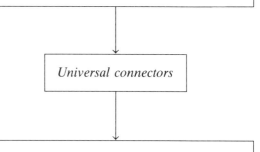

Universal connectors

Computer system and software
- software is multi-tasking – data can be moved around between programs, e.g. in a 'windows' environment
- minimum user decisions
- versatile, high-quality graphics and text presentation

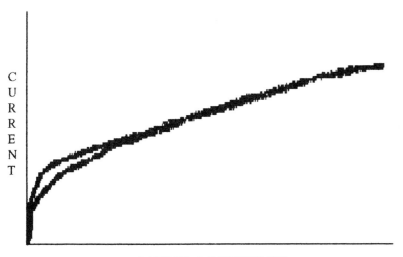

Fig. 8.2 Heating and cooling a bulb. Current v. voltage for a filament bulb

Toolkit Science

This system was devised by Lawrence Rogers and Roy Barton at the University of Leicester. It contains an interface, or 'buffer box', various sensors, software and materials for classroom use which have been produced by teachers. Both analogue data (e.g. temperature readings) and digital data (e.g. from light gates) can be accommodated. It has proved popular despite its lack of sophistication because it is relatively low on cost and high on friendliness. It can be used with Archimedes, Nimbus and BBC computers. Figure 8.2 shows the type of graphical output which can be obtained using this equipment.

Harris blue box sensors

These sensors are well-liked by teachers, despite costing more than some of their competitors and needing to be switched off after use to preserve the battery (Fig. 8.3). There are many sensors in the Harris set, all with similar designs, which means that once a teacher has used one sensor, he or she can cope with all the others.

First Sense and Sensing Science

Data-logging is not confined to secondary schools.

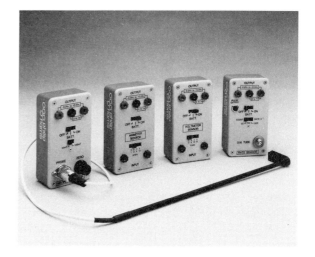

Fig. 8.3 Examples of the range of sensors available

These systems are designed for primary classes and can be used with BBC, Archimedes, Nimbus, IBM and compatible computers (Fig. 8.4).

Softlab

This is a new approach to data-logging software. It is icon-driven, making use of the windows environment.

DATA-LOGGING

Fig. 8.4 Ten-year old pupils monitoring temperature

Fig. 8.5 Using 'Sense and Control'

Fig. 8.6 Using 'LogIT' to monitor an environment containing pond weed

Sense and Control

This is a comprehensive data-logging interface which can accept a wide range of sensors. It can also be used remotely, which means that data can be logged and stored independently of a computer; the data are subsequently transferred to the computer for manipulation, display and printing. In addition to its own software, Sense and Control can be driven by various other packages including Softlab and 'Insight', a data-analysis package developed at Leicester University (Fig. 8.5). Extensive documentation for use with Sense and Control is available from NCET. The interface can connect to Nimbus, BBC, Archimedes, Apple and IBM computers.

LogIT

This is a very compact data-logger which can be used remotely or connected to Nimbus, BBC, Archimedes, Apple and IBM computers (Fig. 8.6). Its sensors are self-identifying and its software comes with optional automatic settings for logging rates. It is very friendly to use; we have seen it used enthusiastically by primary-aged children and postgraduate students alike. For a more detailed review see Scaife (1991).

The Motion Sensor

This is a dedicated data-logger (Fig. 8.7), in that it can only record data about the position of objects in front of it. The package includes software and is designed solely to teach about motion. In our view it does this remarkably well. The system is discussed by Barton and Rogers (1991), who note that 'the factor which grips pupils' interest is that pupils investigate their *own* motion; their personal involvement in the experimenting process is intrinsic'.

DATA-LOGGING

Fig. 8.7 a, b The 'motion sensor': (a) learning to use it   (b) what's happening here?

Key features of the Motion Sensor system are that:

- users produce graphs of their motion in real-time
- they perceive a direct link between the shape of a graph and their motion
- they can predict and test by trying to move so as to match pre-set graphs
- they can see themselves improve in competence while using the sensor

In our view, the emergence of systems like those listed above will begin to convince teachers that data-logging has a rightful place in the Science and Technology curriculum. A list of addresses for data-logging equipment suppliers is given in the Appendix below.

Appendix: Data-logging equipment

- Leicester Toolkit Science Modules: Deltronics, 91 Heol-Y-Parc, Cefneithin, Llanelli, Dyfed, SA14 7DL
- Leicester Toolkit Software: Susan Hemmings, University of Leicester, School of Education, 21 University Road, Leicester, LE1 7RF
- LogIT: Griffin & George, Bishop Meadow Road, Loughborough, Leicestershire, LE11 0RG
- Sense and Control, Educational Electronics, 28 Lake Street, Leighton Buzzard, Beds, LY7 8RX
- First Sense, Philip Harris Educ., Lynn Lane, Shenstone, Lichfield, Staffs, WS14 0EE
- Motion Sensor, Educational Electronics (see above)
- Harris blue box sensors, Philip Harris Educ. (see above)
- Softlab, NCET, Science Park, Warwick University, Coventry, CV4 7EZ
- Sensing Science, NCET (see above)
- Practical Science with Microcomputers (for use with Sense and Control), NCET (see above)
- VELA Users Group, Drs A.R. Clarke and J.K. Jones, Department of Physics, University of Leeds, Leeds LS2 9JT

CHAPTER 9

Interactive video

Introduction: what is interactive video?

This chapter looks at the nature, the use, and the potential of interactive video in Science and Technology education. An interactive video (IV) disc system may typically consist of a computer (including a disc drive and keyboard) which is linked to (or interfaced with) a video disc player and a colour monitor. Often the system can be operated by a light pen, a mouse, a concept keyboard, a bar-code reader or a trackerball, as well as the keyboard. People using IV can control or 'interact' with the system to choose their own sequence of video, sound, text, computer graphics or even still pictures. For example, the Domesday system (developed by the BBC, Philips and Acorn) allows access to a huge amount of data on two laser discs which provide an image of the UK in the 1980s. Users can see maps, moving pictures, photographs and text on the screen. They 'interact' with the system using a trackerball which controls a pointer on the screen. By pointing to different areas on a map, or different parts of a screen menu, users can choose exactly where they want to go and what they see and read.

An example in science was produced by the IVIS (Interactive Video in Schools) project. 'Life and Energy' is a disc produced for middle-year pupils which shows food chains, energy content of foods, and predator–prey relationships. Again, learners can choose from direct instruction, simulations or games using text, still pictures or video. Other examples of IV are described later in the chapter.

In summary, IV can provide a combination of images, sounds and computer-generated text and diagrams which is perhaps unique in a learning situation. IV technology is evolving continually – readers can only keep abreast by studying weekly or monthly publications on, for example, systems which include CD-I (compact disc interactive) and DVI (digital video interactive).

Hardware is certain to change but the educational principles underlying interaction and interactive video are fairly constant.

What is meant by 'interactive'?

The word 'interactive' has several implications which are worth spelling out:

1 *Action/active learning*: The user is involved in some action in using an interactive system. This may involve the use of a mouse or a light pen, pressing keys or using a tracker ball. Different levels of interaction are suggested in the Nebraska Scale (see Table 9.1).
2 *Choice*: In an interactive situation the user has choice over the learning engaged in. The learner can decide which frame to go to, go forward or backwards through the frames one at a time, or freeze a frame.
3 *Control*: Learners have control over their own learning situation (e.g. of the sequence, of the

Table 9.1 Nebraska Scale of Interaction

Level 0: Offering only linear playback facilities (consumer standard), for example video-tape.

Level 1: Using remote control, search, freeze frame, backwards and forwards motion (quickly, slowly and frame by frame), for example laser disc with remote hand-held control or bar-coding control such as the MIST (Modular Investigations in Science and Technology) system.

Level 2: Supporting limited facilities for branching, multiple choice and score-keeping through its microprocessor (e.g. Opensoft, Linkway).

Level 3: Controlled by an external computer, allowing (for example) creation of a scenario (e.g. Ecodisc, Domesday).

Level 4: Using other optical media such as compact discs. Giving rapid replies to a wide range of questions (e.g. through networks to other computers).

---

pace, of the level of difficulty, of the number of repetitions etc.).

4 *The onus* or responsibility for learning rests with the user (largely as a consequence of points 1 to 3). The learner is involved in decision-making and management of learning.

These are all important features of interactivity. They may well apply to other learning situations (e.g. traditional CAL), but they are perhaps most obviously present in IV.

Interactive video (Fig. 9.1) may involve tape or disc and therefore it requires either a disc or a tape player. A tape system is likely to be cheaper but will not have the direct, almost instant access ability of a disc system (cf. the early days of cassette-held programs in school microcomputing).

The educational potential of IV

Superlatives abound in discussions of IV, particularly with reference to data storage. We read of the whole of *Encyclopaedia Britannica* being stored on a single compact disc, and all sorts of predictions about future storage and access from various types of disc. Even with present standards, discs do offer tremendous educational potential. A laser vision disc for example, of the kind used in the Domesday system (see later) can store up to 55 000 still pictures or 37 minutes of video footage on one side. A disc is far more than a video-cassette in this context, largely because of its direct and fast access using digital location of data which can be managed by the computer.

To sum up, the educational potential of IV lies in the following capabilities:

- high storage capacity (for data, pictures etc.)
- highly interactive nature
- motivating power and novelty value
- the possibility for a flexible branching structure

This in turn leads to its potential for tailoring learning to the individual (i.e. for individualised learning of a high degree), allowing the learner almost endless possibilities for repetition and reinforcement. It also allows exploration at greater or lesser depth or practice at a level and for a time to suit the learner's needs. This potential for the individualisation of learning is perhaps the central feature of IV.

In a document early in the history of IV, the Department of Trade and Industry in the UK summed up its potential as a learning medium (DTI, 1985):

> Interactive video combines the visual realism of television with the interactive characteristics of computer based learning. The program demands the participation of the learner by asking questions and presenting him (or her) with choices at appropriate points. Thus the user can explore particular interests, test his own comprehension and reinforce areas of weakness or difficulties, his reactions determining the ensuing pathway through the material.

Uses of interactive video in Science and Technology education

Interactive video is probably the only medium which combines the possibility of all the four paradigms of CAL being used with the same items

INTERACTIVE VIDEO

Fig. 9.1 A CD-ROM system

of hardware and software (see p. 26). In line with the paradigms of Kemmis *et al.* (Table 2.1) we can say that IV can be used in the following modes:

1 *Tutorial mode*: In skills training, in teaching certain concepts, or in teaching a certain body of knowledge. Its training use is quite widespread in commerce and industry (e.g. banks, car industry).
2 *Database mode* (revelatory or emancipatory paradigm): IV can be used almost like an encyclopaedia to allow storage of, and access to, information. Unlike other databases discussed earlier, however, we do not yet have available systems which allow learners to 'write' their own information onto discs – they will be in 'read only' situations.
3 *Surrogate mode*: This will occur when IV is used to simulate certain situations (cf. the revelatory paradigm). Interactive video is often said to provide *surrogate* experiences, for example surrogate walks, surrogate field trips or surrogate travel. It may even, in a science context, provide surrogate experiments of situations which are either too dangerous, too fast or too slow to carry out in real life (cf. computer simulations discussed in Chapter 5).

These modes of use will encourage the 'what if?' questioning and other exploratory or revelatory learning present in database and simulation use generally.

This range of educational uses of IV, within different paradigms, means that it can be used in a variety of different learning and teaching situations. As Atkins (1989) points out, 'the medium does not impose or require a particular style of teaching'. It may lead equally to a whole-class approach or, at the other extreme, small-group or individualised learning. As a learning aid it may be used in a distance learning context, as part of a supported self-study programme or as a central resource in a school or other institution or even a consortium or local authority (for financial reasons this may be the only viable option).

Examples of interactive video

Some of the general points discussed above can be illustrated with specific examples of IV material which are used in schools and colleges.

The Domesday system

This was based on a national project in the UK which aimed to collect information on the country, much as the *Domesday Book* had done many centuries earlier. Like that famous book, the IV system can act as a database with its information stored on two discs: the community and the national disc. Much of the data for the former disc were collected by schools. The discs can be used in Science and Technology education (e.g. in preparing for biology fieldwork, finding geological information, or considering the location of (say) a chemical plant), but has tended to be used more in the humanities curriculum. As well as database mode, the system can be used for surrogate 'walks' and simply for learning by browsing.

Volcanoes

Like the Domesday disc, this system arose from BBC Enterprises (in conjunction with Oxford University Press). It can be used in the form of a superb database with information on volcanoes and plate tectonics allowing open-ended learning. Here are some suggestions of how the disc might be used with students:

- to look at a model of the Earth's structure. Do they think the model is a good one? What evidence is there to support it?
- to find out about how movements inside the Earth can lead to earthquakes and volcanoes
- to compare world maps of volcanic activity with the pattern of plates in the Earth's crust. One piece of film shows the activity at the plate margins.

Ecodisc

This is the third disc which arose from BBC Enterprises, and like the previous two runs on the BBC advanced interactive video system. It is used largely in the surrogate mode, by placing the user in a Nature Reserve in Devon, England. The user can explore, investigate, manage the reserve, or simply take a surrogate walk around the lake and woods. Using Ecodisc will develop many of the important process skills in Science and Technology education including predicting, measuring, evaluating and handling data. A full account of the potential of the disc is provided by McCormick (1987).

Motion: a visual database

This is an IV disc presenting nearly 200 short film sequences of a wide range of examples of motion. It is an excellent resource for a range of abilities, allowing extremely detailed exploration of motion for older or more able pupils or more qualitative study and discussion of motion in earlier years (e.g. why is it best for a car to 'crumple' when it collides with a wall? Why wear seat belts and crash helmets?). A full review of this resource, which is much more than a straight database, is given in Scaife and Wellington (1989).

The Newcastle Interactive Learning Project

This is a project which has produced discs on Energy, Radioactivity, Risk, Probability and Statistics. The discs on energy and radioactivity

INTERACTIVE VIDEO

Fig. 9.2 a, b Stills from the interactive videodisc, Motion: a Visual Database

largely involve simulation activities which fully exploit the potential of the medium (e.g. on siting a nuclear power station). Users can explore the Periodic Table using two surrogate guides (two teenage girls). A full review of the project materials is given by Wellington (1990a).

Problems and cautionary notes

It is easy to become excited about IV as a learning medium, especially during a demonstration of its potential away from the classroom. Some cautionary notes are well worth making.

Firstly, the cost of hardware and software for schools and colleges is high and looks likely to remain so, in relative terms. Secondly, the cost and mental effort required in developing and producing good educational software is enormous. This means that, even if an institution can afford an IV system, there is unlikely to be even a steady trickle of software which can be used on it. This is coupled to the third problem, which is the rapidly evolving nature of the technology. Who knows where the technology for interactive video learning will be in three, let alone five, years from now? This may in turn make software producers reluctant to invest time and money. On a more optimistic note, there may come a time when critical mass is reached when there is enough hardware around to entice effort into the production of software, and thereby enough good software to encourage the purchase of hardware. When this critical mass or chain reaction situation is likely to occur, and with which kind of IV technology, is somewhat difficult to predict.

Finally, there is a paucity of research on the use and evaluation of IV in learning situations. It may be wise at this stage to treat some of the wilder claims for IV with healthy scepticism. What can be stated with some certainty is that the use of interactive video in Science and Technology education should not be a substitute (a surrogate or virtual reality) for work in the real world, and that however sophisticated the system, it will never replace the teacher – it may however necessitate adaptation of the teacher's role.

CHAPTER 10

Word-processing and desk-top publishing

Introduction: WP and DTP – where do they fit?

This chapter examines the use of word-processing (WP) and desk-top publishing (DTP) in Science and Technology education, and considers the value of these tools in both areas. At first glance both WP and DTP fit most easily into the emancipatory paradigm (see p. 26). Both are in a sense labour-saving and may well take away some of the drudgery or 'inauthentic labour' from certain aspects of Science and Technology education. But in another sense they save no labour and certainly no time at all in practice. Their achievement is to create new ways of learning, new forms of labour, and new possibilities in Science and Technology education. We will look at some of those possibilities, and one or two of the issues involved in the use of WP and DTP – not least, the question of what counts as authentic and inauthentic labour.

Word-processing in Science and Technology education

Most pupils and students still write by hand. Write-ups of experiments, evaluations of a project, design briefs and so on are probably more likely to be handwritten than typed on a keyboard. Most people would argue that writing with a pen or a pencil is an essential skill and should be preserved. Few would disagree. A small minority would go further and suggest that the use of a keyboard to write words, a computer to process and store them, and a printer to print them will actually hinder or stunt the development of handwriting and even writing generally. We have heard this point of view expressed at meetings, courses and in discussion. Will the keyboard oust the pen?

In fact there is little or no evidence to support this understandable fear. In a project where schools and homes were virtually saturated with computers (the Apple Classroom of Tomorrow project, or ACOT) it seemed that pupils' writing was enriched and certainly increased through the use of computers. Pen and pencil were still used. The nature of writing was also changed (see Wellington, 1990b, for a summary of the ACOT project).

We would not like to pretend that the debate is a simple one. It certainly remains a live issue and one which merits far more school-based research in different contexts. But it is now already clear that the use of WP will provide the following enrichment and benefit:

1 Pupils are given the opportunity to draft and re-draft their own work much more readily. This is well known, and all those who have used a WP system will have experienced it. It seems to affect different users in different ways. Some are much more inclined to actually make a start on a piece of writing (arguably the hardest part of the process), knowing full well that it can easily be changed or edited. Some are actually much more inclined to keep going, just to get

their thoughts down onto paper or the screen, knowing that they can easily be re-drafted. This aspect of WP is often said to enhance the writing of so-called lower-ability students – but our suspicion is that it affects writers at all levels.

2. Pupils are able to collaborate (work co-operatively) on a piece of writing much more easily with a computer system than with pen and paper. Partnership in writing is encouraged. This occurs for perhaps two main reasons: firstly, the writing is up there on the screen for all partners to see. This enables them more easily to take an equal share in it. Secondly, the writing is actually physically done by a shared keyboard, there on the desk or bench. It often does not 'belong' to one person more than another as, say, a pen does.

3. Marking of work done on a WP system can be so much more painless. Again, this applies equally at all levels of education and writing. Writers are far more inclined to seek feedback and critical comment if they know that alteration, addition and editing are relatively simple. This is again said to apply especially to those most likely to make spelling or grammatical mistakes, which is certainly true – the use of WP does remove the need for marks and corrections all over a script. But it can have an influence on people's writing attitudes and habits at all levels.

4. The final product of a piece of writing can be so much better through the use of WP and DTP, as we will see later. This can produce a positive feedback loop, in turn influencing the earlier stages.

5. Finally, writing done with a WP system can be easily stored and exchanged. On the one hand this may encourage malpractice with exam coursework, though we know of no cases. Its positive effect is to allow a person or a group to stop writing at a convenient point and take it up again more easily later.

These are some of the main general points connected with the use of WP in Science and Technology education. A further aspect of interest in the use of WP which we have witnessed in schools is that teachers who were otherwise reluctant about using IT were often willing users of WP and DTP. Their activities tended to focus on the production of teaching materials such as worksheets, assignment sheets and tests, as well as general course and departmental documentation. The seemingly anomalous situation of a teacher owning several boxes of floppy discs while never using a computer in a lesson seems not to be uncommon. One possible interpretation of this is that as WP is an extension of something familiar, namely typing, and the user is firmly in control of events, the technology is therefore relatively non-threatening to the user. It is accepted by IT sceptics because the pay-off outweighs the threat.

When used by teachers to produce teaching materials and professional documentation, WP can contribute to departmental teamwork. It is much easier to circulate draft copies to colleagues and accommodate their suggested changes when the document is stored on disc. In this sense, IT can be seen as a potential aid to consultative and management processes in schools.

Issues in the use of word-processing

There is a distinct shortage of written accounts on the use of WP in Science and Technology education. It would be so valuable if teachers would report their own experiences or small-scale research projects in this area. One notable exception is given by a primary headteacher who reported the use of WP to write up the practical science investigations of small groups of children (Clough, 1987). He reports the following effects:

- greater attention to detail in the investigation itself as a result of using WP to write it up
- more 'efficient' writing up, in terms of quantity and quality
- improved note-taking during practical work
- increased motivation, particularly of 'less academically able' pupils, in carrying out investigations resulting largely from the more attractive presentation

This teacher's account is largely qualitative and based on a small but nonetheless valuable sample. It raises several issues which are worthy of further research.

A more general study was carried out by Peacock (1988). He analysed the marking of work in English of a respectable sample of pupils – some of the work had been word-processed and printed, some had been hand-written. In summary, his general conclusions were as follows:

- word-processed work is graded more generously by teachers in English Language (more so than English Literature)
- this is especially true of the lower quality work
- at higher levels the 'glamour' effect of WP work is 'almost entirely absent'
- the less directly that work relates to 'an external referent' (e.g. a novel, a poem), the greater the enhancing effect of WP

Peacock goes on to ask: 'How should exam boards react to these findings?' He suggests various possibilities from banning WP course work to removing any restrictions. Examiners and moderators would then need to be trained to make them aware of the factors reported in the research.

Peacock's work is interesting from a Science/Technology perspective, particularly with regard to the way that teachers react to, and mark, word-processed written work. They need to be aware of the effects pointed out by Peacock in their attitude to WP and DTP work in Science and Technology.

Further studies of the kind undertaken by Clough and Peacock would be particularly valuable in illuminating the effects (both positive and negative) of WP in students' written work in Science and Technology. This is an area which truly does merit further study and debate on the important issues involved. Two questions in particular are of central importance:

1 Should pupils be taught keyboard skills, using the ubiquitous QWERTY layout at an early age? One of the barriers to the use of WP (and databases, see Chapter 7) has been pupils' slowness in typing in text. Should typing be taught to all pupils as a matter of course?
2 A second barrier to use, as with many aspects of IT, is access to a computer system. What will be the effect of increased use of portable/laptop computers on the incidence and use of WP? Should they be introduced into Science and Technology lessons on a wide scale, especially as pupils increasingly acquire their own laptops (Peacock and Breese, 1990)?

Desk-top publishing

Word-processing programs are designed to manage text made up of letters, numbers and other symbols such as those found on typewriters. The layout of the text can be altered in several useful, but strictly limited ways. Desk-top publishing (DTP) programs are more flexible. Text layout can become a design issue. DTP can accommodate line drawings and sprites (pictures made of groups of pixels; a pixel is a small area of the screen) as well as data from other programs. The DTP user might utilise newspaper format in columns, text flowing round graphics, attractive data display such as a three-dimensional pie-chart, text enhancements including a variety of fonts, headings and borders (Fig. 10.1). DTP programs can also operate as simple word-processors, which is not as retrograde as it appears because printing can be much quicker in this mode of use.

What additional possibilities does DTP offer in Science and Technology education?

1 Pupils' presentation of work can be greatly enhanced. The quality of presentation, a tacit issue in assessing written material, can become an explicit component through which work is judged. Many teachers (including the authors) may have mixed views about the educational value of this *per se*. Few would dispute, however, that DTP has a place in the technology curriculum in graphics courses. One teacher told us: 'I think of the computer as a complete God-send, especially in the graphics course – this course is centred around the computer.'

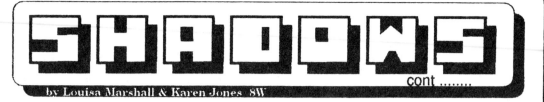

by Louisa Marshall & Karen Jones 8W

In this part of the experiment we wanted to see if the distance between the lamp and the pot made any difference to the width of the shadow.

We placed the lamp 3 cm above the table and positioned the pot at each centimetre distance from 5cm to 10 cm from the bulb. The width of the shadow was then measured at both 10 cm and 20cm along its length.

Distance of pot away from bulb	Width of shadow at 10cm	20cm
5cm	7.5cm	11.0cm
6cm	6.5cm	10.5cm
7cm	6.5cm	10.0cm
8cm	6.0cm	9.5cm
9cm	5.5cm	8.0cm
10cm	5.3cm	7.5cm

We have shown these findings in the form of a graph on the following page and it can be seen that the change in distance between the bulb and the pot results in a corresponding change in the width of the shadows.

In the next part of this experiment we wanted to find out what effect altering the angle of the light bulb would have on the length of the shadow.

cont ....

Fig. 10.1 Students' DTP report

WORD-PROCESSING AND DESK-TOP PUBLISHING

Fig. 10.2 Example of graphics application which can be used in desk-top publishing

2 DTP can be effective in group work, where one or two students take responsibility for using DTP to present the work of the group.
3 The ability to include graphics allows DTP to be used to present plans, design specifications, project reports and effective data display (Fig. 10.1). A primary function of each of these documents is to communicate effectively with the reader. Presentation is fundamental in this process.
4 The quality of teaching materials can be improved, if teachers can find time to learn how to use DTP. If the use of word-processing is any indication, it seems that many teachers will make time.
5 The presentation of the school 'image' can be enhanced. We are arguing less that this *should* be a part of teaching than that it has already *become* so, and that IT in the form of DTP can play a significant role.

Associated programs

With the development of DTP, 'document processing' is a much broader art and science than was 'word-processing'. Programs have been developed to be dedicated to particular aspects of document processing such as data display or sprite management (Fig. 10.2). Work can be done in these programs and can then be 'exported' into a DTP program so that whole documents can be brought together, much as a newspaper or magazine is a collection of individually produced items of text and graphics.

The example in Figure 10.1 is part of an investigation by two year-eight students. The presentation of the DTP report was the students' own work, and formed part of the overall project.

CHAPTER 11

Spreadsheets

Introduction

A spreadsheet is a program which deals with information in the form of a table, with rows and columns. The rows are often given numbers and the columns are given letters so that any particular cell or element of the table can be identified, for example as C5 or J2. Data can be changed or linked to other data by specifying which elements are to be changed and what the nature of the change is to be. An example would be to multiply every number in column D by the corresponding number in column E and put the results in column F. If column D contained data about speeds and column E contained data about time intervals, then column F would represent distances. This facility for data manipulations is a key characteristic of spreadsheet programs. Spreadsheets can also be used for sorting and displaying stored data and for creating and manipulating mathematical models.

Some manipulation processes can be carried out by using either a spreadsheet or a database. Which one is preferable? A rough guide might be that databases tend to offer more powerful search and sort options while spreadsheets tend to be easier to inspect, update and edit.

Uses for spreadsheets in teaching

A simple form of spreadsheet may be dedicated to accepting only one type of data, for example, the food consumed by every pupil in a class in one day. The data are manipulated in prescribed ways, with the aim of communicating information in a pre-programmed format. Although restricted in use, such programs are accessible to students and they can be a gentle introduction to the use of more versatile spreadsheets.

The more flexible and powerful the spreadsheet, the more the user is required to make decisions about the way the data are stored and manipulated. These decisions may include the user having to classify the data into groups and identifying the relationships between groups of data.

Three spreadsheets which have proved popular with teachers in our acquaintance are, in order of power, 'Grasshopper' for BBC and Nimbus computers, 'Pipedream' III and IV for Archimedes and 'Excel' for Nimbus and IBM-compatible computers. In addition, the Cellular Modelling System has the structure of a spreadsheet, and while it has very limited capacity for data, it is nevertheless a powerful tool for computer modelling. It is described in more detail in Chapter 6.

Any activity in Science or Technology which involves students looking at or building up tables of information might be considered as a candidate for spreadsheet use. Examples might include the following: design of buildings to minimise fuel consumption (see Chamberlain and Vincent, 1989); environmental decision-making, such as the siting of a new chemical plant; investigating a predator–prey relationship; correlating experimental data, such as current, voltage, resistance and power.

It has been argued (see Goodfellow, 1990) that spreadsheets occupy a middle ground in terms of the demands made on students by software, between the passive receiver (when using simulations) and the total creator (when using modelling applications). Goodfellow claims that 'if the purpose of an experiment is to change the model of reality in the child's mind, it is best changed gradually by the constant review of that model by the child itself'. He suggests that spreadsheets are appropriate in this respect because (a) results can be readily processed, (b) results can be meaningfully presented and (c) the processing medium (the spreadsheet), can be easily accessed by the majority of children. He illustrates these points with the work of nine and ten year olds using 'Grasshopper'.

The above argument is that the use of spreadsheets can enhance students' learning in Science and Technology (and we would add Humanities too). What is the balance between pay-off and cost to the teacher? The cost, in its broader sense, involves acquiring and learning to use the software, at least so as to remain one step ahead of the students (or only one step behind, as we heard wryly remarked in a staffroom). It also involves building IT into schemes of work and lesson plans, and booking a computer at the right time, in the right place. Finally the students must be shown how to use the software. The pay-offs from using spreadsheets must include the potentiality for improved student learning and motivation, otherwise they would have no place in the classroom. Additional advantages, which may be less obvious could include the following:

- *Flexible learning*: students can work independently and at their own pace
- *Working cooperatively in groups*
- *Teacher–pupil relations*: pupils will notice and value the fact that the teacher has bothered to introduce a new and interesting activity
- *Improved teacher competence in IT*: teachers may find uses for spreadsheets in their professional work other than with their classes. Departmental accounting and stocktaking, and collation of assessment schedules and examination results are two likely areas
- *Increased teacher confidence in IT*: teachers who have hitherto been wary of computers may find that simple spreadsheets are much easier to use than they feared

CHAPTER 12

Computers in technology laboratories

Introduction

This chapter focuses on aspects of IT which, through custom and practice in schools, have been generally associated with practical work in Technology. This includes control technology, control of experiments, process control, computer-aided design and manufacture (CAD and CAM), music technology and microelectronics. The list is not claimed to be exhaustive; it reflects a range of activities which have been identified both in the literature and in our case studies as significant in terms of the use of IT.

Control

Switching things on and off is the essence of digital electronics and control technology, as Sparkes (1989) succinctly pointed out, so there is little wonder that microcomputers are now being used to teach these topics.

Control technology with young children might involve the use of a programmable robot, such as 'Bigtrak' or 'Roamer'. In order to send Bigtrak, which is a robotic vehicle, on a pre-determined path children have to discover how to control certain key operations: motor on or off, motor direction forward or backward, how many forward steps are needed to move 'Bigtrak' by 1 metre, how many turn steps are needed to rotate by a right angle. Robotic toys like 'Bigtrak' are fun and they make for a positive and exciting introduction to programming for control. They can suffer from physical limitations however, such as different step numbers needed on different surfaces (for example, wooden or carpeted floors) in order to rotate by a right angle. Such limitations can preoccupy users to the detriment of their learning. One remedy is to program an object to move on the computer screen, the best known example of which is probably LOGO. LOGO is a serious programming language, not a toy or a game, despite being associated with turtles. For a general account of the purposes and uses of LOGO in teaching and learning see Papert (1980). LOGO uses commands which are meant to be transparent to the programmer, such as 'Forward 100' to move the turtle forward 100 steps and right 50 to rotate it by 50 steps. (These commands can be abbreviated to speed up the programming.) The turtle is a screen-based icon. Children using LOGO rapidly acquire control over the turtle and may progress quickly to programming with procedures and recursion.

The screen turtle allows children to become proficient at programming for control without distractions caused by friction, uneven floors and so on. Having gained confidence in the language they can return to the problem of controlling the movement of real-life objects such as a 'buggy' or a Valiant Turtle (Fig. 12.1a and b). This can be achieved by connecting an interface to one of the computer's output ports. The program statements cause voltage pulses to be sent to the interface, which then converts the pulses into a suitable form

Fig. 12.1 a 'Roamer' in use: a controllable floor vehicle

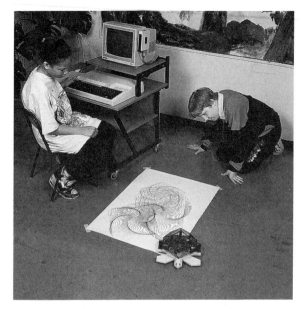

Fig. 12.1 b The 'Valiant Turtle' in action

to drive the vehicle. Typically, the commands have the same form as those described above for 'Bigtrak'.

If the vehicle is a buggy made from a deconstructable kit such as 'Lego', possibilities now arise for controlling other types of models, such as a railway-crossing barrier or a crane. And if lights and buzzers are available the models can be further developed.

Control technology in practice

A description of some of the practical issues involved in introducing control technology into the curriculum of a primary school is reported by Buckle (1988). The sessions were timetabled with a group of 28 10- and 11-year olds to occupy one afternoon per week. Continuous blocks of time were preferred to several separate periods so that design processes could be followed through. The first role for the teacher was to provide some core knowledge about the new equipment (chiefly the interface), that the class was going to use. This was accomplished by demonstrating some functions of the interface and drawing up a reference chart, to be displayed subsequently on a classroom wall. Groups of pupils were then directed to try writing some short programs to control various output devices. The programs were tested on the computer and de-bugged as necessary, with the teacher in the role of consultant. In subsequent sessions the pupils constructed 'Lego' models for which they wrote control programs. Buckle observed that in due course, all of the groups (some of which included children with special educational needs) were working at a fairly high level and with great enjoyment. Buckle's description closely matches our own observations of control technology in the primary curriculum (Fig. 12.2).

Some implications for teachers

Through its very nature, control technology shifts the teacher's role from the centre of the stage, except for occasions such as the introduction of core skills and knowledge. For the majority of the

COMPUTERS IN TECHNOLOGY LABORATORIES 85

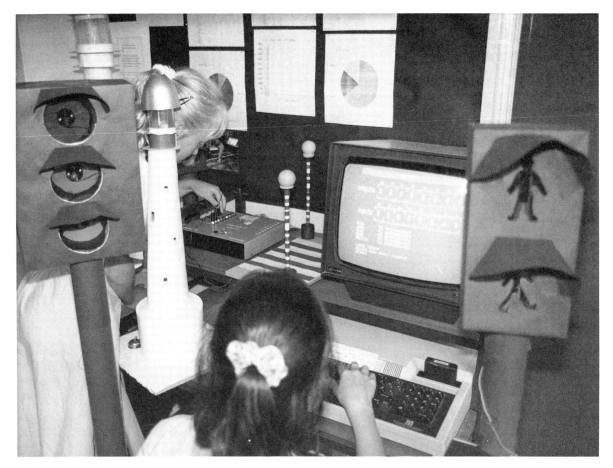

Fig. 12.2 Pupils controlling traffic lights

time however, the teacher acts as consultant, project manager and chief technician. For teachers who equate the centre-stage position with classroom control, this shift of role may be intolerable, at least at first.

One particular aspect of the teacher's role is likely to be significantly altered from its traditional form, namely that of the assessor. In project work, pupils are engaged in the process of solving problems, which in turn means that they will be exploring needs, trying out designs, attempting to put plans into reality and evaluating their actions and ideas. In the course of all this they will inevitably make mistakes. Traditional forms of assessment penalise mistakes and direct students towards 'correct', closed outcomes. By contrast, the teacher as assessor of project work will recognise the pedagogical value of 'mistakes' and also of the learning environment in which mistakes can be tolerated. The teacher-assessor will also share the responsibility for evaluation with the students.

In another role, that of chief technician, the teacher will be expected to know as much as possible about the computer, the interface and the software. It is desirable for the teacher to gain confidence and some skills in trouble-shooting, ideally through in-service training, though this tends to be rare.

INFORMATION TECHNOLOGY IN SCIENCE AND TECHNOLOGY EDUCATION

Controlling experiments

The use of computers to record experimental data has been discussed in Chapter 8. There are some situations in which the computer could also be used to control experiments (Fig. 12.3). One example is when an experiment runs for a long period. This was the case in an experiment involving one of the present authors, in which the behaviour of a metal specimen was investigated under conditions of on–off stretching and on–off heating. The computer took the required measurements and then controlled the action of heaters and pumps to set the prescribed loading conditions. This was repeated for several days. A further example, described by Khan (1985: 65) used a computer to control a laser and a motor-driven camera to take measurements of the girth of a tree under conditions which would have been undesirable for a human operator.

Fig. 12.3 A compact control and monitoring system: 'LogIT'

Fig. 12.4 'Autocad' in use

Process control

The quality of peripheral hardware now available to schools is high enough to allow realistic simulation of industrial process control. A working model of, for example, a conveyor system or a heating system, can be interfaced to the computer and controlled with software. Students investigate how to operate components such as motors and solenoids and they then learn how the process as a whole is controlled. Typically, this involves 'tuning' the output devices by introducing delays into software procedures. Teachers who are fortunate enough to be able to make links with manufacturers who use process control systems may choose to use related examples in their teaching. If they are really fortunate, they may find manufacturers who are willing to donate equipment!

CAD and CAM

With the introduction into schools of 16- and 32-bit computers with their associated graphics capabilities, computer-aided design (CAD) has become established in technology teaching. In conjunction with a plotter, packages such as LINCAD and AUTOCAD (Fig. 12.4) can be used by students for high-quality draughting and graphic communication. One technology teacher in our locality has used the CAD components of a CAD-CAM package principally with design students who wanted to train in architecture. Although the CAM part of the package could link the computer to a lathe, this option had not been used. Another school had developed useful links with the local hosiery industry. The design department was in the process of developing a control program to drive a knitting machine using patterns produced on a computer.

An approach that integrates CAD-CAM with other aspects of design technology is described by Taverner (1990). Using a program for the BBC/Master micros called 'Vehicle Design', pupils are able, according to Taverner, to design the profile of a car, the properties of which are then modelled by the computer. Pupils are presented with calculated values for the principal design parameters, including fuel economy and drag coefficient. They then use this information to refine their vehicle's body shape. If they choose, say, to add spoilers, they can investigate the effect in a screen simulation of a wind-tunnel test. Having optimised their profiles, students can use their printouts to make vacuum-formed models of the cars. Taverner concludes that the software 'genuinely integrates IT with designing and making'.

IT and Music Technology

In the not-too-distant past, the main and perhaps only link between music and design technology in schools arose because students sometimes opted to design and make guitars. Now, with the availability of fast, high-resolution sound-sampling techniques, sounds can be stored in the form of sets of digital data. Since computers are ideal for manipulating and displaying digital data, a new form of music has been born. Not for the first time in the history of Music Technology, a new device has produced new styles. Moreover, computer-literate students have found electronic music accessible, both to listen to and to create. The medium has also attracted interest from teachers, especially as costs for sampling systems have fallen sharply since they first became available. We know of a school in our locality with an Atari computer, 'Cubase' software, a keyboard, a multi-track recorder and a 12-channel studio. There is undoubted scope with such a system for creative liaison between school Music and Design departments – a prospect which could be equally exciting for students and for teachers.

Microelectronics

According to Sparkes (1989), the systems approach to electronics is here to stay. The use of modules, as opposed to discrete components, has given students the opportunity for simple, quick construction of electronic systems. These systems tend to work reliably and can be incorporated into interesting small-scale project work – factors which contribute to a high level of motivation in their users.

One of the more widely used modular kits is the Microelectronics For All (MFA) system. This includes a range of inputs, outputs, logic gates, counters and memories. Students will probably begin with an investigation of the characteristics of logic gates. They can progress to the design and construction of systems such as burglar alarms, and then to some elementary programming. Finally, the modules can be connected to a computer or a music module, or they can be made to control a buggy.

MFA kits have been used with children aged 10–14 years. More versatile modular systems are available for older students.

Many teachers, both experienced and new to the profession, approach the topic of microelectronics with trepidation and fear of the unknown. They should not despair, for although microelectronics is one of the newer elements in the Science and Technology curriculum, it is also one of the least conceptually demanding. It is an area where a little hands-on experience goes a long way to establishing confidence.

A list of software for use in the Technology laboratory is given in the Appendix.

Appendix: Software in the Technology laboratory

- Logotron LOGO, Longman, Dales Brewery, Gywdir Street, Cambridge, CB1 2LJ
- LINCAD, Linear Graphics Ltd, Unit 39, Mochdre Industrial Estate, Newtown, Powys, SY16 4LE
- 'Pro Artisan', Clares Micro Supplies, 98 Middle-wich Road, Rudheath, Northwich, Cheshire, CW9 7DA
- 'Vehicle Design', Heineman Software, The Argent Centre, 60 Frederick St., Birmingham, B1 3HS
- CUBASE, Steinberg, Evenlode Soundworks, The Studio, Church Street, Stonesfield, Oxford, OX7 2PS

SECTION C

ISSUES AND POLICY

CHAPTER 13

Issues to be addressed

Case studies

To highlight some of the issues that are arising with the use of IT in the curriculum, we visited a number of schools and colleges covering the age ranges: primary, middle, secondary and tertiary. Four institutions, one for each age range, were chosen as case studies. The choice of schools was made on the basis of the range of activities taking place within them and the number of teachers involved in IT in Science and Technology. Time was spent in interviewing appropriate staff to discover what they saw as the current, pressing issues in IT. Views were obtained from management, departmental heads and classroom teachers.

While such a survey cannot be regarded in any formal sense as representative of schools at large, we believe that the views we obtained provide some indication of the broader pattern of IT policy and practice in Science and Technology education. These views are outlined in this chapter.

The evaluation of CAL: questions, claims and evidence

How far has the use of IT in Science and Technology education actually produced advance or improvement? This question, or a question in a similar form, is often asked – particularly by cynics. Unfortunately the question is a very problematic one. It depends on the meaning of the terms 'advance', 'improvement' or terms of a similar nature used by a cynical questioner. In addition it would be extremely difficult to obtain a satisfactory answer to the question by any sort of empirical analysis or experiment. How could an experiment be designed to show that (say) CAL had produced an improvement in Science education?

These introductory remarks are merely to introduce a note of caution and wariness into claims made for IT use and evaluations of it. However, this is not to suggest that no claims can be made or evaluations carried out by teachers or researchers – we simply suggest that claims made and questions posed should be examined carefully to see if they are even capable of being supported or refuted. The claims made by Papert for the use of LOGO and other forms of CAL provide a perfect example – as Hannon and Wooler point out (1985) they are not open to falsification (Popper's term) or, by the same token, confirmation.

Modest claims, with evaluation to support them, have been made and are worth considering. The literature is full of them and only a few examples are singled out below.

Chatterton (1985) for example, studied the use of CAL in Science classrooms in a range of schools using a Systematic Classroom Analysis Notation (SCAN). As a result of his observations he suggests that 'the use of CAL can produce a qualitative improvement in the learning environment'. In particular CAL will promote:

- more individual or small-group work

- more open-ended questions
- more task-oriented pupil–pupil discussion
- more vocalisation by the pupils of their understanding of the topic being covered
- greater opportunity for the pupils to develop and test their own hypotheses

These claims are well worth further reflection and perhaps investigation by teachers or school-based researchers. Similarly, Olson (1988) makes claims for the use of IT not through the systematic observation of Chatterton but as a result of eight case studies. Olson suggests that computer use affects both teachers and pupils. He suggests that teachers often use computers to express things about their own practice – to make a personal statement. As a result teachers' use of IT leads them to:

- re-think what they teach (cf. Sendov's second wave, see p. 21)
- re-think their role and their methods i.e. how they teach
- re-consider their values

Olson's findings in this area are supported by our own case studies. As one Head of Science at a comprehensive school put it:

> IT use should lead to a shift in the way that teaching and learning take place ... the way that pupils work. For example, if pupils become adept at using spreadsheets from the middle school right through to Year 11, they really don't want to go back to pencil and paper. IT can give kids a new dimension in science ... it can make some aspects of science more accessible.

This quotation links with Olson's second point, that pupils' learning is affected by the use of CAL. He suggests that pupils are encouraged to 'confront their own preconceptions in a critical way'. If true, this claim is particularly appropriate for learning in Science with the vast evidence now accumulated on the power or even obstinacy of pupils' preconceptions or alternative frameworks. Olson also claims that through IT use pupils learn to work with each other in new and unfamiliar ways. This is important for the teacher. It follows, as Olson puts it, that teachers need to accommodate their routines and perhaps risk their own classroom influence in order to assimilate IT use.

Our final comment here stems from Olson's ideas. The successful introduction of IT into Science and Technology education must be a process of accommodation and assimilation. This is the adaptation which teachers must undergo. Some of the barriers to this process are considered below.

Access to IT: management and resource issues

> ... the main barrier to IT use is that teachers cannot always rely on the equipment being there when they need it – therefore they don't plan it into their lessons.

This was the comment made by the deputy head of a local comprehensive school who is also a Technology teacher. He felt that the school policy should be to place IT equipment where it is likely to be used – to be 'hard-nosed about it'.

Access to IT was not felt to be the major barrier by every teacher (see below). However, the issue of management and deployment of resources in IT did surface frequently in the case studies, and has also featured in previously published studies. Some of those studies are worth considering briefly before turning to the main issues involved in our own case studies. Bliss, Chandra and Cox (1986) for example, examined the factors which influenced the introduction of computers into one school, at a time when the 'Micros in School Programme' was very much in its first phase or wave. They classified seven different types of teacher as follows: the favourable, the critical, the worried, the unfavourable, the antagonistic, the indifferent and the uninitiated. These types are still recognisable, although many of the feelings may well inhere in one and the same person – how many of us do oscillate between different feelings and attitudes towards IT at different times and in different situations? Bliss *et al.* went on to paint portraits of two teachers at opposite ends of a spectrum which now looks rather dated but does

ISSUES TO BE ADDRESSED

raise some issues which are still relevant. Perhaps the main point is that the *management of people* and their attitudes is as important as the *management of resources*. Indeed, the people of an institution such as a school are its most expensive and its most valuable resource.

A similar study, though in the primary sector, was made by Ellam and Wellington (1986). Their study focused on a 10 per cent random sample of primary schools in one large local education authority. The research identified a number of positive factors which facilitated IT use in the primary school. The most important factors can be classed under the label of management – of both people and resources. Positive factors included:

- a supportive headteacher
- the provision of adequate non-contact time for teachers to work alongside other members of staff and to develop appropriate IT use
- a teacher with responsibility for developing IT in the school (e.g. by cataloguing and organising software; by organising INSET)
- a computer policy which has been jointly developed by staff
- a variety of software of different categories (as classified in Chapter 2), though not an excessive amount (i.e. quality not quantity)
- facilities and accessories, such as trolleys, to make the computer more readily available to the busy classroom teacher and to reduce setting-up time

These were all factors which seemed to lead to a 'diffusion' of IT use within the school.

In secondary schools the successful management and diffusion of IT may well be more difficult to achieve. This may be due partly to the increased size, but it must also be partly attributable to the highly subject-based curriculum in secondary education. The strong subject emphasis has led to the development of territories and 'territorial imperatives' within IT. Certain subject departments (often Maths, Science and Technology) have traditionally moved further in IT than others and may wish to preserve their lead. This inevitably presents a management problem. This situation has become more complicated in England and Wales with the publication of the National Curriculum Statutory Orders for Science and Technology. IT within the National Curriculum could be organised in schools as a specific set of IT courses, as a set of themes to be covered, say, by the Science and Design Technology departments, or as a cross-curriculum activity which permeates the whole school. Many teachers would see the first option as a retrograde step unless it were to be accompanied by one of the others. The second option is organisationally simpler than the third, but it excludes the majority of teachers in the school. And the third option would present some interesting challenges when students' IT learning came to be assessed! The issue will not be pursued further here, as we have little research evidence to illuminate it – but it is worth signalling as a potential barrier to the diffusion of IT.

Finally, a useful NCET publication has highlighted 14 organisational conditions for the successful development of IT within a school which can be used as a framework for those involved in managing IT. The conditions are shown in Table 13.1.

The subject of IT resource provision is, of course, highly political. The British Government

Table 13.1 Organisational conditions for successful development of IT

1 Identification of subject IT leaders
2 Establishment of IT sub-committee
3 Formulation of school IT policy
4 Coordinated school focused INSET
5 IT coordinator on curriculum committee
6 Participative academic organisation
7 Leadership based on expertise
8 Head well informed and supportive of IT
9 High level of teacher cooperation
10 Teachers feel recognised and appreciated
11 High level of parental involvement
12 Close liaison with local authority
13 Adequate technician support
14 Understanding and support of governors

Source: Adapted from MITS *Project Newsletter* 4, April 1990, NCET.

has made much of its policy in the 1980s of establishing 'computers in every school', a policy which, taken literally, has been very successful. As teachers are quick to point out though, hardware provision is but one aspect of the mosaic of resources necessary for the effective implementation of IT education. (Training is examined below.)

The numbers of computers in schools have steadily risen, but so have expectations and aspirations. The Parliamentary Office of Science and Technology (POST, 1991), recently recommended that one computer for 12 children was needed in order to meet the requirements of the National Curriculum. Few of us can think of a school where such a ratio is approached. In order to reach this level of resource nationally, the National Association of Inspectors and Advisers for Computers in Education (NAACE) estimates that an additional 230 000 extra computers are needed!

The financial implications of the inclusion of IT in the National Curriculum are thus enormous (see for example Hart, 1990). Who will grasp this nettle? Parents and local businesses are already, according to Lewis (1991), contributing as much as local government, namely between 20 and 40 per cent of the total bill. This was borne out by a survey of 450 teachers reported by Cole (1992). The survey was carried out at the 1992 BETT conference in London and indicated that 38 per cent of all schools rely on funding from parent–teacher associations to buy computers. Some people see this as evidence of proper parental involvement in schooling and of strong school–industry linking. Others argue that it will widen the gap between the best and worst equipped schools, with resource levels in schools in prosperous areas far exceeding those of schools in poorer areas.

What of the schools themselves, which now have financial control over virtually their entire staffing, resourcing and maintenance budgets? In the words of one headteacher:

> We are certainly not planning any expansion of IT resources or staffing in the future because of the financial position. There is no obvious successor to TVEI when current hardware begins to fail.

Other barriers to the diffusion of IT use

The case studies were particularly valuable in illuminating the real barriers which teachers face at all levels in using and developing IT in Science and Technology education (and we would suggest, in other curriculum areas). The main barriers are as follows:

Time

Teachers need, above all else, time to learn how to use IT. Time is needed to become familiar with hardware and software. As one teacher put it:

> Staff have to invest time in learning IT before IT can actually save them time.

In many cases IT will not save them time anyway – it will simply change the way they work.

Time is also needed to gradually integrate the use of IT into the classroom. It may take the course of a whole school year before a program or package becomes 'built in' to the department curriculum, a module, or a scheme of work. This is something which headteachers, in the view of at least one teacher interviewed, do not fully appreciate.

Finally, staff need time to 'find out what software is available' – this is seen as a bigger barrier to development of IT use than a lack of software itself.

Attitude

Many teachers felt that 'attitude' was a key barrier (cf. Bliss *et al.*, 1986). Lack of 'ease' with IT, or even fear were mentioned. On a positive note one teacher felt that 'being prepared to make mistakes, in an atmosphere where risk-taking is accepted' are essential.

These are both issues which can be addressed to some extent by in-service training, which is discussed below.

Deployment of resources

A further issue in secondary schools, which has been reported previously, concerns the siting and

ISSUES TO BE ADDRESSED

deployment of IT resources. In a study reported in 1989 evidence on computer use in secondary schools suggested that placing computers in a 'computer room', and in addition networking those computers, did place a restriction on the use of IT across the school (Wellington, 1989a). Deploying resources in this way may well prevent more flexible use of IT in, for example, Science and Technology. As one teacher said:

> A networked computer room is useful if the whole class is working on the same program. But as computers have found their way into labs there has been a move away from the mass use of simulations.
>
> The big disadvantage of the network is that only one person really knows how to manage it. The network does seem to meet the needs of the Business Studies staff – but the network has posed problems for staff who don't know how to use it properly.

Our impression is that even now the majority of secondary schools still rely almost exclusively on the 'computer room' as 'the place to do computing'. Until this centralised domain is dissolved by some kind of computer reformation, micros in secondary schools will not be widely and flexibly used in all subject areas, including Science and Technology.

Staff development in IT: the bootstrap model and the learning cycle

How have Science and Technology teachers learned about IT? In response to questions on this issue, staff from a range of specialisms almost invariably respond that they were self-taught. Some teachers added that occasional help had come from other sources including students, the local university or college, teacher colleagues and initial training courses. What was striking was the haphazard nature by which IT expertise had been acquired by these teachers. Our view is that this situation is common throughout the education system.

In-service training provision was not regarded as a significant source of IT learning for teachers. It was the subject of some criticism:

> Our experience of INSET at LEA level has been off-putting. Colleagues have tried it and been treated like kids.

According to another teacher:

> Anything that came through on an IT heading tends to be irrelevant to my uses.

Design advisers had proved more helpful for this teacher.

Some teachers who were actively involved in IT in their schools seemed unaware that they could look to their local authority for support:

> I'm not sure what they could do. I've never really enquired.

The greatest time investment on local authority INSET reported to us by one of the teachers was 'four or five days in about 10 years' which, the teacher added, was reasonable. It seems that some Science and Technology staff are resigned largely to teaching themselves about IT. As it happens the INSET time mentioned is similar to the figure of '1 day with an advisory teacher every $3\frac{1}{2}$ years' quoted in the POST report (1991) as the outcome of a £90 million DES initiative in 1988 to employ 650 IT advisory teachers over three years.

Initial training seems to have provided little IT expertise for the teachers interviewed, but they were all experienced and had completed their initial training some years, if not decades, earlier. Future entrants to the profession will be better served in this respect, if for no other reason than that accredited courses of initial teacher education in England and Wales must now contain substantial elements of IT training.

Our enquiries revealed that the 'normal' route to IT expertise in Science and Technology was through on-the-job self-help. This depends very much on the inclination and enthusiasm of the individual, which inevitably leads to some staff giving a low priority to IT or even opting out altogether. One head of department commented:

> I taught myself, basically. Now I try to teach colleagues but they're mostly a bit resistant. I've tried formal internal INSET but it's not been very productive.

A headteacher put it this way:

> It demands a fair amount of determination. It's the bootstrap model.

The issues of staff development and INSET are closely bound up with that of resource management, i.e. the availability of hardware and software. As one teacher suggested, there is a kind of 'learning cycle' in IT which requires three elements: INSET, staff development time, and availability of software and hardware. These three elements are needed in order for progress in IT to take place.

A worrying picture emerged with regard to the issue of software availability. The case of a science teacher who was responsible for managing part of the school's IT equipment was reported to us. The total allowance for the year did not run to the cost of one site licensed program:

> I'll have to try and scrounge some money. And the database – that would have been over £500 for a site license. I brought a single user version and . . . just . . .

This teacher's dilemma is anything but unique. On the one hand is the commitment to improving the quality of IT teaching and learning in the school. On the other hand are the financial constraints and the moral and legal implications of trading in 'hot' merchandise. Underfunding in IT is in danger of turning highly-principled teachers into criminals!

Gender issues

There is a large literature on the gender issue in relation to IT use. However, an excellent summary of the key points in this area is provided by Hoyles (1988).

The list below is our attempt to paraphrase those key points which have particular bearing on IT use in Science and Technology education. Research evidence including recent work by Harrison and Hay (1991), suggests that boys are more likely to:

- have a computer at home
- be interested, and motivated towards, computers
- have a positive attitude towards computers
- have networks of friends interested in computers
- take over computers during study periods and break times at school (the girls not wanting to compete with them)
- volunteer to do things on the computer for the teacher
- secure a greater share of resources and teacher attention than girls
- gain access to the newest, most powerful computers

These findings all have important implications for classroom teachers in Science and Technology (and indeed other areas). In particular they need to be borne in mind in considering classroom management and control when IT is being used. This is especially true when access to equipment is limited and competition is likely. Teachers should reflect on, at the very least, the following questions:

1 If computer work at home (e.g. writing-up), is suggested or even positively rewarded, is this fair and just?
2 If volunteers to demonstrate IT equipment are sought, is attention paid to both sexes?
3 When pupils first collect IT equipment in the lesson do teachers ensure that both sexes can gain access to the resources? Are certain pupils left with the older equipment? Should teachers positively discriminate in allocating resources?
4 Where IT access is scarce (i.e. in almost every real situation), is *time* (e.g. use of a word-processing system) and *access* (e.g. to data-logging equipment) shared equally?

Progression in IT

One final issue which surfaced in the case studies, and seems to be a perennial one, is the problem of ensuring progression in the use of IT. How can teachers monitor progression and development in the use of IT? Can it ever be controlled and guided in the same way as (say) progression in Physical Education, Music or Science – especially as there

ISSUES TO BE ADDRESSED

are so many factors outside the teacher's control, including the use of home computers?

More fundamentally, what counts as progression in the use of IT? Does it mean that students should gradually learn to use more complex and sophisticated programs, i.e. progression in depth? Or does progression involve a gradual broadening of a student's horizons and awareness of IT, i.e. increased breadth? And will not the rapid rate of development in IT make a nonsense of any attempt to develop depth and breadth? A student in compulsory schooling will live through at least 11 years of technological change. Students are faced with technological progress which proceeds independently of their own education. In addition, students who leave school this year are unlikely to have encountered any IT which is more sophisticated than that which five-year olds will meet in their primary years – due simply to the sheer pace of IT development.

These are some of the general issues involved in considering progression. In our case studies, the issue surfaced particularly in relation to the primary–secondary transition. Secondary teachers for example, were concerned that programs they were using had already been introduced in primary school (e.g. 'Grass' and 'Grasshopper' spreadsheet and database). Teachers felt an urgent need 'to learn more about the IT capabilities of the children coming to school'. As one head teacher put it: 'We need to know where the children are at.' (Hints are given here perhaps of a constructivist approach to IT learning.) The same head wanted to develop a computer certificate for the children about to leave primary and enter the comprehensive. This would have two effects: 'better curricular liaison, staff working together ... and we would know what they have a mastery of'.

Other ways of improving the IT transition have no doubt been tried. However, the issue of progression within a school still remains.

The non-statutory guidance for IT in the National Curriculum (DES, 1990) lists five aspects of progression in IT; these involve students moving through problems of increasing complexity and unfamiliarity with increased skill and independence. It is also stated that progression will involve the use of more sophisticated software, though progression in terms of hardware, with its implications for continued investment in resources, is conspicuously absent from the list.

The non-statutory guidance provides a model for schools to conduct an audit of IT coverage, continuity and progression. Five strands of IT are identified, which can be traced through from Key Stage 1 to Key Stage 4, covering the ages 5–16 years. The five strands are 'communicating information', 'handling information', 'modelling', 'measurement and control', and 'applications and effects'. As an illustration, the following extracts from statements of attainment demonstrate progression in the 'measurement and control' strand.

- *Key Stage 1*: pupils should be able to talk about ways in which toys and domestic appliances can respond to signals or commands
- *Key Stage 2*: pupils should be able to develop a set of commands to control the movement of a screen image or robot
- *Key Stage 3*: pupils should understand that data can be logged over long or short periods or at a distance
- *Key Stage 4*: pupils should be able to explain how they have used feedback in a monitor and control system

The difficulty of conducting an IT audit is considerable and it represents a different kind of challenge for schools from ensuring continuity and progression in, say, Science. The audit must involve every department in a school, since IT is supposed to permeate the whole curriculum. It must include IT in the feeder schools and, ideally, in the fed schools. It must take account not only of the current hardware and software in the cluster of schools but also of the next generation of these materials. And, finally, in spite of the shifting ground of sudden and sometimes unpredictable initiatives and developments in IT education, teachers must do their best to ensure smooth and systematic progression by their pupils.

CHAPTER 14

Issues for the future

Introduction

Chapter 13 discussed some of the issues concerning IT use in Science and Technology education which have already emerged and which surfaced most commonly in our own case studies of IT in different phases of education. This final chapter speculates on the future of IT use in education. Speculation of this kind is notoriously risky and the history of IT predictions (what could be called 'Yesterday's Tomorrows': Steele and Wellington, 1985b, p. 196) is littered with howlers such as the 1940s forecast by British experts that five large computers would be enough to serve Britain's needs! Healthy scepticism is thus required.

New developments in software

Already we have seen significant shifts in the nature of educational software since the early 1980s (Table 14.1). There has been a move from the use of skill-and-drill, content laden software towards content-free, open-ended software which affords much greater freedom to the user. Software has become student-centred rather than teacher-led. Programs which encourage and even depend on collaborative use have become increasingly common, in contrast to the individual, linear-path software of a decade ago. What could be called the 'battery-hen' mode of computer use (with individuals pecking away at keyboards) is less common, although it can still be seen in many secondary school computer rooms. Software was often age-specific. Now the possibilities of the same programs being used at different ages and stages of education is much more widespread (which brings us back to the earlier problematic question of what counts as progression in IT).

These are all changes which teachers have lived through in the 1980s and 1990s. Is it possible to predict with any certainty the changes in waiting? Obviously, more powerful computers will bring improved graphics and sound capabilities – these changes are occurring constantly. In addition, educational software is increasingly likely to involve a range of media – multimedia – rather than simply the traditional VDU in front of the user.

One of the major changes which has been discussed for some years now (O'Shea and Self, 1983)

Table 14.1 Trends in educational software

Task-specific	→ Content-free
Linear	→ Exploratory
Teacher-led	→ Student-centred
Directed	→ Open-ended
Individual	→ Collaborative work (e.g. collaborative writing)
Age-specific	→ Fostering continuity (not age-specific)
VDU-focused	→ Multimedia
Textual, obscure	→ Iconic, intuitive

Source: Adapted from a talk given by Steve Heppell at the 'Resource' IT Conference, November 1989.

is the introduction of some form of artificial intelligence (AI) into educational software. What constitutes AI is a complex field (see O'Shea and Eisenstadt, 1984, for an early introduction). However, it is worth mentioning here because of its implications for Science and Technology education. AI software is apparently in its infancy. The introduction of 'intelligence' into software will bring about a 'quantum leap'. One development will involve expert systems. These are programs which store in a huge database expertise which has been elicited from humans and then fed into the system. There are problems over how this expertise is elicited and whether all kinds of expert knowledge (for example, professional 'wisdom') is capable of elicitation, but essentially the 'expert system' contains a broad base of knowledge on a certain subject (e.g. human ailments, oil prospecting). The expert system, like a database, can be interrogated by the user. But in addition the system will make deductions and draw inferences from the information fed into it by using rules which are stored in the system and themselves elicited from experts. Thus an expert system will not only be able to offer advice and opinion to the user (e.g. on a human ailment which it has diagnosed and how it should be treated) but will also be able to show its own reasoning leading to any conclusions it has drawn. This latter feature is essential in allowing the user to examine and question any conclusion or suggestion – it seems to be an extremely sound educational principle if expert systems are ever to become a feature of IT in education.

A dilemma, however, with an expert system is that as its base of data and algorithms (the 'rules' by which it forms decisions) expands, its operation necessarily becomes all the more complex. In other words, the wiser the expert system, the more complex it becomes to interrogate by the 'lay person'.

It is open to speculation as to how these systems may prove valuable in Science and Technology education – perhaps in identification of species or chemicals?; perhaps in design applications of (say) structures?; perhaps in the diagnosis of faults in a design or product (cf. medical diagnosis)?

Trends in the technology itself

It is conventional to talk of different generations of computers, starting from the *valve* machines of the 1940s, through the *transistor-based computers* of the late 1950s, to the *integrated circuits* of the 1960s and 1970s, and then the *very large-scale integration* (VLSI) of the fourth generation. The speculation now concerns the so-called *fifth-generation computers*, which have been under development since the mid-1980s. Up to this generation, computers had been built around a single central processing unit (CPU), operating with a single sequence of instructions (a program). The new generation involves devices which are 'massively parallel', with many processors working together. This phase, it is said, will include computers capable of responding to natural language. Fifth-generation computers will also be intelligent and capable of making decisions (i.e. they will have artificial intelligence). Predictions on the future of the fifth generation are many and varied – however, if (as seems to be a common prediction) new systems can understand natural language, the implications for education will be interesting to say the least, particularly if systems can cope with the spoken word. The user interface between learner and computer will be transformed. Where will the QWERTY keyboard, the overlay keyboard, the mouse, and other input devices be left? Will the possibility of voice input change the way that computers can be used in practical subjects like Science and Technology?

In addition, future microcomputer systems are certain to possess:

- vastly increased internal processing power and speed
- greater internal memory capacity
- increased capacity in external storage media linked to the system
- greatly increased portability due to these and other advances
- ability to communicate with each other in common operating languages

The possible implications of these changes are also

ISSUES FOR THE FUTURE

open to speculation. One obvious growth in education, however, lies in the spread of portable devices such as *palm-top*, *notebook* and *laptop computers*. Some research and evaluation work has been done on the use of laptops, although little on a large scale. Problems seem to have been experienced with machines which are perhaps not suitable for technical reasons (e.g. batteries losing power quickly and thus work being lost; lack of a disc drive, let alone a hard disc). One published study describes loss of work as 'commonplace' (Peacock and Breese, 1990). These are problems which need to be solved if groups of children are not to be given a poor attitude to the use of portable machines. The ideal situation is to have machines with an advanced power source such as a continuously re-charging solar battery, a casual storage medium with a higher capacity than current magnetic floppy discs and lighter-weight bulk storage in place of current hard discs. In addition we might expect input to laptops to be via a pen and an active pad which recognises handwriting and user-defined shorthand, and helps the user with drawings. Currently medium–capacity laptops cost in excess of £1000, but this price will surely come down. One other perceived disadvantage is the negative effect of laptops on collaborative work if the screen cannot be seen by a small group working round it, as is the case with many portables. This is another technical problem which needs examining.

Benefits of portable computer power are evident in reported studies. Pupils seem to adopt a more flexible approach to writing (e.g. in drafting and re-drafting). It seems that writing with laptops (as with other computer systems) will develop a new attitude amongst teachers and pupils to the process of writing. Their use also seems to lead to readier and more flexible access. In areas such as Science and Technology education, laptops seem to have made little impact as yet, but they should not be forgotten in developing new ways of working in the future.

The development of AI will surely be accompanied by increased sophistication in input and output devices such as sensors and senso-mechanisms, allowing commercial production of intelligent robotic devices. This evolution of hardware will have direct influence on the use of IT in Science and Technology for measurement and control. Some of these developments may result in significant changes in the way we teach, say, mechanics or heating and conservation. We already have a small foretaste, with the arrival of the Motion Sensor (see Chapter 8). Teachers who have integrated this device into their mechanics schemes have reported new – and more effective – forms of teaching, resulting from the new technology (Barton and Rogers, 1991).

Another technical development, much-heralded in the early 1990s, has been multimedia. Also described as the 'collision of computing and television', multimedia systems combine the traditional text and graphics of computers with audio and video information. The principal storage medium in multimedia systems is the CD-ROM, not least because the CD format is well established in the audio market. The CDI system, developed by Phillips, can be used either for leisure and entertainment or for teaching and learning, depending on the choice of software.

There is, however, a major limitation in the capability of disc-based multimedia systems to manage moving pictures. This results from the physiology of human vision; in order for a sequence of still pictures to appear to the eye as a smooth 'movie' without flicker, the stills must appear and disappear quickly – at least ten times per second and preferably many more. But a full-screen high resolution still image can require up to 1 Mb of memory. At ten such screen changes per second, a CD-ROM could store about one minute of movie! Floppy discs are even slower. One software reviewer pointed out that it took about 3.5 minutes for a 25 MHz 386 PC to generate a single 320 by 200 frame (Schofield, 1992).

Strategies used by manufacturers to overcome this demand for high speed information include compressing the images and using small sections of the screen for movies. Limited memory saving can be achieved without loss of visual quality. Until this memory flow problem can be solved, the effectiveness of multimedia systems in classrooms will be severely limited. Teachers may well find the larger IV disk format more useful, especially if combined with a fast access system such as a bar-code reader.

The home–school interface

One of the issues which has already emerged in IT use and seems certain to grow in importance concerns the home–school interface in the use of computers, particularly with the spread of portable IT. A forward-looking article by Hannon and Wooler in 1985 (in Wellington, ed, 1985: 84–95) discussed some of the questions raised by the use of IT resources in the home for educational purposes, such as the issues of equal opportunities and parental involvement which may be highly dependent on social class and economic circumstances.

Since that time there has been relatively little research on the home–school interface in IT, with the notable exception of the Apple Classroom of Tomorrow (ACOT) project. One of the experiments within the ACOT project involved giving a third-grade class (eight- to nine-year olds) unlimited access to a computer both at school and at home – a policy described as 'total saturation'. Pupils were given their own Apple machine to take home and keep for the duration of the project. What were the effects of such immediate and unlimited access? The findings were complex (a summary and contacts for further information are given in Wellington, 1990b) but one of the obvious effects was a marked improvement in keyboard skills. In addition, the writing skills of the pupils was said to show great improvement, both in quality and quantity, partly as a result of their improved keyboard skills. In other areas of the curriculum pupils were able to do work at school and take it home on a floppy disc to finish, or vice versa. This happened, for example, with LOGO work which pupils may have started at home and gone on to finish at school. Home–school liaison has not been without its problems, however. Home use appeared to centre largely on either word-processing or skill-and-drill software use. The former did create classroom management problems, with pupils lining up to print out their homework from disc. Perhaps the answer here is total printer saturation, or, alternatively, for teachers to accept homework on discs rather than on paper. The latter was considered to be useful in some ways, but perhaps not educationally highly desirable.

Parents have been involved with the ACOT research. When the Apples were ready to go home, 'family members' were given 10 hours of training. In this way the project has been able to explore the issue of parental involvement in IT, an important issue for the future. Eventually, a wide range of publications will stem from the project.

A new 'pedagogy of information'

Gwyn (1988), in a speculative discussion on the future of IT in education, talks of the need to develop a new 'pedagogy of information'. Although guilty of technological determinism in talking rather glibly of the 'Age of Information' and the uncontrollability of technological change, he does pose some interesting questions on the future of teaching and learning. He asks, for example, questions about the kind of knowledge which pupils will need to retain in their heads when factual knowledge becomes a 'shirt-pocket, transportable commodity':

- what learning experiences will be appropriate for pupils who bring their own portable computing power to the classroom which is equivalent to current mainframe computers?
- how much control will individual teachers be able to exert over the nature of 'the learning experiences available to pupils'?
- finally, with the spread of IT and especially communication devices and networks, will society still need school as an institution?

Gwyn's discussion is speculation of the pure Alvin Toffler variety (Toffler is the American author of *Future Shock and the Third Wave*, Bantam Books, 1980) but it is nonetheless thought provoking.

Our *personal view* is that the school will sur-

ISSUES FOR THE FUTURE

vive (for a variety of reasons, not all concerned with education) but the curriculum will surely be influenced and adapted as a result of IT development and of course many other changes. The roles of student and teacher will evolve, perhaps even faster, attitudes to learning and assessment will change, approaches to lessons and lesson planning will be modified. Some possibilities for the 'new curriculum' which may develop in the future – partly as a result of IT and partly for other reasons – are summarised in Table 14.2. We can only hope that the development of IT use in Science and Technology education will contribute to some of these changes.

Table 14.2 Is this the 'new' curriculum?: possible trends in teaching and learning

Aspect	Conventional approach	The 'New' Learning
Approach	Content-driven	Process-driven
Focus	Teacher-led	Student-centred
Teacher role	Expert	Fellow learner/facilitator
Emphasis	Knowing that	Knowing how
Student activity	Working alone	Work in small groups
Ethos	Competitive	Collaborative
Student role	Passive/receptive	Active/generative
Lessons	Programmed/pre-planned	Flexible opportunity
Topic	Imposed	Negotiated
Mistakes	Should not be made	Are to be learned from

Bibliography

Alvey Committee (1982) *A Programme for Advanced Information Technology*. London: HMSO.

Atkins, M. (1989) Visual stimulus: the exciting world of interactive video lessons. *Times Educational Supplement*, 5 May.

Atkins, M. and Blissett, G. (1989) Learning activities and interactive videodisc: an exploratory study. *British Journal of Educational Technology*, 20(1): 47–56.

Baker, C. (1984) A critical examination of the effect of the microcomputer on the curriculum. In C. Terry (ed.) *Using Microcomputers in Schools*. London: Croom Helm.

Barton, R. and Rogers, L. (1991) The computer as an aid to practical science – Studying motion with a computer. *Journal of Computer Assisted Learning*, 7: 104–12.

Bleach, P. (1986) *Computer Use in Primary Education*. Reading: University of Reading.

Blease, D. (1986) *Evaluating Educational Software*. London: Croom Helm.

Bliss, J., Chandra, P. and Cox, M. (1986) The introduction of computers into a school. *Computer Education*, 10(1): 49–54.

Brown, J.S. (1985) Process versus product: a perspective on tools for communal and informal electronic learning. In M. Chen and W. Paisley (eds) *Children and Microcomputers*. California: Sage.

Buckle, R. (1988) Raising barriers with control technology. *Education*, 3(13): 8–12.

Cerych, L. (1985) Problems arising from the use of new technologies in education. *European Journal of Education*, 20: 2–3.

Chamberlain, P. and Vincent, R. (1989) Using a spreadsheet to analyse the thermal performance of a model house. *School Science Review*, 71: 39.

Chandler, D. (1984) *Young Learners and the Microcomputers*. Milton Keynes: Open University Press.

Chatterton, J. (1985) Evaluating CAL in the classroom. In I. Reid and J. Rushton (eds) *Teachers, Computers in the Classroom*. Manchester University Press.

Cloke, C. (1988) Information technology in science. *Computer Education*, 60: 25–6.

Clough, D. (1987) Word processing in the classroom and science education. *Primary Science Review*, 5: 4–5.

Cole, G. (1992) Number cruncher. *Times Educational Supplement*, 12 June, p. 41.

Dawes, M. (1983) The microcomputer and the curriculum. *Journal of Curriculum Studies*, 15(2): 207–14.

Department of Education and Science (1986) *Statistical Bulletin, 18/86: Results of the Survey of Microcomputers in Schools*. London: DES.

Department of Education and Science (1988) *Science for Ages 5 to 16: Proposals of the Secretary of State for Education and Science and the Secretary of State for Wales* (August). London: HMSO.

Department of Education and Science (1989) *Science in the National Curriculum*. London: HMSO.

Department of Education and Science (1990) *Technology in the National Curriculum*. London: HMSO.

Department of Trade and Industry (1985) *What is Interactive Video?* Information Sheet. London: DTI.

Dieuzeide, H. (1987) Computers and education: the French experience. *Prospects*, 17: 531–7.

Ellam, N. and Wellington, J.J. (1986) *Computers in the Primary Curriculum*. Sheffield: University of Sheffield.

Frost, R. (1991) *IT in Science Blue Book*. North London Science Centre, 62–66 Highbury Grove, London, N5 2AD.

Gagné, R.M. (1985) *The Conditions of Learning*. New York: Holt, Rinehart and Winston.

Goodfellow, T. (1990) Spreadsheets: powerful tools in science education. *School Science Review*, 71(257): 47.

Gray, T. (1984) Versatile focus. *Times Educational Supplement*, 19 November, pp. 51–2.

Gwyn, R. (1988) Towards a pedagogy of information. In R. Ennals, R. Gwyn and L. Zdravchev (eds) *Information Technology and Education*. Chichester: Ellis Horwood.

Hannon, P. and Wooler, S. (1985) Psychology and educational computing. In J.J. Wellington (ed.) *Children, Computers and the Curriculum*. London: Harper and Row.

Harrison, C. and Hay, D. (1991) Hi-technique of manual skills. *The Guardian*, 7 February.

Hart R. (1990) *We can't Afford the National Curriculum*. Sheffield Education Computing Centre, 20 Union Road, Sheffield, S11 9EF.

Hoyles, C. (ed.) (1988) *Girls and Computers*, No.34. London: Bedford Way Papers.

Jacobson, E. (1987) Microcomputers: opportunities and challenges to reshape the content and method of teaching maths and science. *Prospects*, 17(3): 407–16.

Kahn, B. (1985) *Computers in Science*. Cambridge: Cambridge University Press.

Kemmis, S., Atkin, M. and Wright, S. (1977) *How do Students Learn?* Occasional Paper No.5. CARE: University of East Anglia.

Lewis P. (1991) Parents asked to plug the funding gap. *Times Educational Supplement*, 7 June.

Macdonald, G. and Wellington, J.J. (1989) Computers in the classroom: *TES* survey. *Times Educational Supplement*, 17 March, B22–B27.

Mackintosh, I. (1986) *Sunrise Europe: The Dynamics of Information Technology*. Oxford: Basil Blackwell.

Maddison, A. (1982) *Microcomputers and the Classroom*. London: Hodder and Stoughton.

Mashiter, J. (1988) Interactive video in science. *School Science Review*, 69(248): 446–50.

McCormick, S. (1987) Ecodisc: An ecological visual simulation. *Journal of Biological Education*, 21(3): 175–80.

Moore, J.L. and Thomas, F.H. (1983) Computer simulations of experiments: a valuable alternative to traditional laboratory work for secondary school science teaching. *School Science Review*, 229(64): 641–55.

National Curriculum Council (1989a) *Technology 5–16 in the National Curriculum: A Report to the Secretary of State for Education and Science on the Statutory Consultation for Attainment Targets and Programmes of Study in Technology*. York: NCC.

National Curriculum Council (1989b) *Science: Non-Statutory Guidance*. York: NCC.

Ogborn, J. (1984) *Dynamic Modelling System*. London: Longmans Microcomputer Software.

Ogborn, J. (1986) *Cellular Modelling System*. London: Longmans Microcomputer Software.

Ogborn, J. (1990) A future for modelling in science education. *Journal of Computer Assisted Learning*, 6: 103–12.

Olson, J. (1988) *Schoolworlds/Microworlds: Computers and the Culture of the Classroom*. Oxford: Pergamon.

O'Shea, T. (1985) *The Learning Machine*. London: Broadcasting Support Services for the BBC.

O'Shea, T. and Eisenstadt, M. (eds) (1984) *Artificial Intelligence: Tools, Techniques and Applications*. New York: Harper and Row.

O'Shea, T. and Self, J. (1983) *Learning and Teaching with Computers*. Brighton: Harvester Press.

Owen, M. (1991) *IT in Science Teacher Development Project*. Bangor: University of Wales.

Papert, S. (1980) *Mindstorms: Children, Computers and Powerful Ideas*. New York: Basic Books/Brighton: Harvester.

Parliamentary Office of Science and Technology (POST) (1991) *Technologies for Teaching, Vols 1 and 2*. Little Smith Street, London SW1P: POST.

Peacock, M. (1988) Handwriting versus word processed print. *Journal of Computer Assisted Learning*, 4: 162–72.

Peacock, M. and Breese, C. (1990) Pupils with portable writing machines. *Educational Review*, 42(1): 41–56

Robins, K. and Webster, F. (1989) *The Technical Fix: Education, Computers and Industry*. London: Macmillan.

Rogers, L. (1990) IT in Science in the National Curriculum. *Journal of Computer Assisted Learning*, 6: 246–54.

Rollinson, H.R. (1989) Interactive video in Earth Sciences education. *Teaching Earth Sciences*, 14(1): 15–16.

Roszak, T. (1986) *The Cult of Information*. Cambridge: Lutterworth Press.

Royal Society (1990) *Interactive Video and the Teaching of Science*. London: Royal Society.

Rushby, N. (1979) *An Introduction to Educational Computing*. London: Croom Helm.

Ryle, G. (1949) *The Concept of Mind*. London: Hutchinson.

Scaife, J. (1991) Facts fingered: review of the LogIT

BIBLIOGRAPHY

data-logger. *Times Educational Supplement*, 7 June, p. 53.

Scaife, J. and Wellington, J.J. (1989) Caught in action: the motion video-disc. *Times Educational Supplement*, 1 September.

Schofield, J. (1992) God's-eye view finder. *The Guardian*, 18 June, p. 31.

Self, J. (1984) *Microcomputers in Education*. Brighton: Harvester.

Sendov, B. (1986) The second wave: problems of computer education. In R. Ennals, R. Gwyn and L. Zdravchev (eds) *Information Technology and Education*. Chichester: Ellis Horwood.

Sewell, D. (1990) *New Tools For New Minds: A Cognitive Perspective on the Use of Computers with Young Children*. Hemel Hempstead: Harvester Wheatsheaf.

Smart, L. (1988) The database as a catalyst. *Journal of Computer Assisted Learning*, 4: 140–9.

Sparkes, R.A. (1985), *The BBC Microcomputer in Science Teaching*. London: Hutchinson.

Sparkes, R.A. (1989) Information technology in science education. *School Science Review*, 71(245): 25–31.

Spavold, J. (1989) Children and databases. *Journal of Computer Assisted Learning*, 5: 145–60.

Steele, R. and Wellington, J.J. (1985a) Hardware and humans. *Times Educational Supplement*, 26 April.

Steele, R. and Wellington, J.J. (1985b) *Computers and Communication*. Glasgow: Blackie.

Thatcher, M. (1982) Introduction. In *Department of Trade and Industry, Micros in Schools Scheme*. London: HMSO.

Taverner, D. (1990) Body building. *Times Educational Supplement*, 15 June, p. 13.

Turkle, S. (1984) *The Second Self: Computers and the Human Spirit*. London: Granada.

Underwood, J. and Underwood, G. (1990) *Computers and Learning*. Oxford: Basil Blackwell.

Weizenbaum, J. (1984) *Computer Power and Human Reason*. Harmondsworth: Penguin.

Wellington, J.J. (1984) Computers across the curriculum: the needs in teacher training. *Journal of Further and Higher Education*, 8(3): 46–53.

Wellington, J.J. (1985) *Children, Computers and the Curriculum*. London: Harper and Row.

Wellington, J.J. (1988) *Policies and Trends in IT and Education*. Lancaster: ESRC-ITE Programme, University of Lancaster.

Wellington, J.J. (1989a) *Education for Employment: The Place of Information Technology*. Windsor: NFER-Nelson.

Wellington, J.J. (ed.) (1989b) *Skills and Processes in Science Education*. London: Routledge.

Wellington, J.J. (1990a) Right on cue: the Newcastle Interactive 'Video Project'. *Times Educational Supplement*, 10 August.

Wellington, J.J. (1990b) California dreaming: the ACOT Project. *Times Educational Supplement*, 23 March.

Index

artificial intelligence (AI), 24, 100
authentic labour, 23
automaticity, 38–9

CAD, 87
CAL, 18, 34, 37–41, 91
CAM, 87
CD-ROM, 101
Cellular modelling system, 52
Computer Studies, 16
conjectural paradigm, 25
control technology, 83–7
controlling experiments, 86

databases, 34, 57–60
datafiles, 59–60
data-logging, 34, 61–8
desk-top publishing (DTP), 34, 75–80
Domesday system, 72
drill-and-practice, 37–9
Dynamic modelling system, 52

Ecodisc, 72
emancipatory paradigm, 25
evaluation of CAL, 91–2
expert systems, 100

fifth generation computers, 100

gender, 96

home-school interface, 94, 102

inauthentic labour, 23
INSET, 95–6
instructional paradigm, 24, 37
interactive video (IV), 69–74

laptops, 101
Logo, 54, 83, 88

MEP, 15
MESU, 17
microelectronics, 87–8
modelling, 51–5
Motion Sensor, 66–8

multimedia, 99, 101
music technology, 87

National Curriculum, 31
Nebraska Scale of Interaction, 70

paradigms, 24, 26
pixels, 77
portable computers, 101
process control, 87
programmed learning, 26, 37
progression, 96–7

revelatory paradigm, 24

simulations, 24, 34, 45–9
skill-and-drill, 37–41
spreadsheets, 81–2
sprites, 77
surrogate experiences, 71

TVEI, 16

WIMP, 36
word processing (WP), 34, 75–80